PROJECTS AND
COMPLEXITY

PROJECTS AND
COMPLEXITY

EDITED BY
FRANCESCO VARANINI
WALTER GINEVRI

CRC Press
Taylor & Francis Group
Boca Raton London New York

CRC Press is an imprint of the
Taylor & Francis Group, an **informa** business
AN AUERBACH BOOK

CRC Press
Taylor & Francis Group
6000 Broken Sound Parkway NW, Suite 300
Boca Raton, FL 33487-2742

First issued in paperback 2019

ISBN-13: 978-1-4665-0279-6 (hbk)
ISBN-13: 978-0-367-38141-7 (pbk)

This book contains information obtained from authentic and highly regarded sources. Reasonable efforts have been made to publish reliable data and information, but the author and publisher cannot assume responsibility for the validity of all materials or the consequences of their use. The authors and publishers have attempted to trace the copyright holders of all material reproduced in this publication and apologize to copyright holders if permission to publish in this form has not been obtained. If any copyright material has not been acknowledged please write and let us know so we may rectify in any future reprint.

Library of Congress Cataloging-in-Publication Data

Projects and complexity / editors, Francesco Varanini, Walter Ginevri.
    p. cm.
  Includes bibliographical references and index.
  ISBN 978-1-4665-0279-6 (hbk. : alk. paper)
  1. Project management. I. Varanini, Francesco. II. Ginevri, Walter.

HD69.P75P7634 2012
658.4'04--dc23
                                           2012009653

Visit the Taylor & Francis Web site at
http://www.taylorandfrancis.com

and the CRC Press Web site at
http://www.crcpress.com

The English edition of this book is dedicated to Carlo Notari

for his unwavering support, and generous and endless

inspiration in helping to transform an idea into this book.

Contents

Foreword

This is an important book, for three principal reasons.

First, it addresses complexity, an area that is becoming increasingly important in the management of projects. There has been a developing literature on this subject, and over the last five years or so the profile of project-based complexity has been on a significantly accelerating trajectory.

Second, this book looks at complex projects through a number of interesting lenses, and in a cultural context that is quite different from many academic accounts. A markedly different sociological, philosophical, and cognitive perspective has been applied to the subject area, exposing new and novel viewpoints that add some thought-provoking insights to an already stimulating debate.

Third, emerging project management seeks to unravel a number of misconceptions about the field of project-based management, and attempts to document an "emerging" model of project management that embraces elements and incorporates concepts that take us in a distinctly different direction from the traditional and historically accepted "plan, then execute" project paradigm. In time it may become one of the cornerstones of what we may want to call "Project Management 2.0."

The contributions offered here look at project management in such a way that your curiosity is awakened, and Walter Ginevri and his team of contributors, researchers, and colleagues are to be commended for this. I personally became aware of this material through a meeting with one of the contributors at a research conference. After that meeting, and a spirited discussion about the complexity that pervades and surrounds the modern project domain, Walter personally sent me a copy of the original text, in Italian.

Although I was very pleased to receive this kind and well-meant gift, unfortunately, the content was not very accessible to me (it is probably fair to say that my command of the Italian language would be challenged by the need to order a cup of coffee). My only Italian colleague at the time was not particularly entranced about carrying out a personal and essentially "bespoke" translation of all 300 pages of the original edition, so I was delighted to learn that progress was being made toward offering a published English translation. We have now arrived at that moment.

Inevitably, in the translation from the original Italian, I suspect that there are places where some of the subtlety of meaning is marginally either lost or distorted, and at times therefore, this is not a relaxing or a straightforward read. However, it has often been said that nothing worthwhile is ever acquired easily, and the rewards for those who persevere include an opportunity to look at the issues documented here in a quite different light.

So, what is this book trying to tell us? Vitally, one of the key messages, and one that has been emerging from a number of different directions and research perspectives, is that project-based management is much more than the execution of an agreed-upon and defined plan. No one is suggesting that planning be abandoned, but this text is keen to stress that projects are uncertain, and that the turbulent environments in which they are executed add to that uncertainty.

This means that a range of emerging factors within the project management domain are pushed into prominence, and the content of emerging project management seeks to address these from a number of different perspectives. There will be some issues raised here with which the "traditional" project manager (if there is such a thing) will be profoundly uncomfortable. Looking at project management through a humanistic and philosophical lens is revealing, and often disturbing, but it is an important way of unraveling and dissecting elements of project-based work that are often taken for granted.

This does not mean that some of the more "traditional" elements within project-based management are ignored; there are sections in this text that deal with WBS, stakeholders, risk management, leadership, and with other mainstream components of project management. However, different philosophical and cultural viewpoints are adopted, and we can all learn from adopting different perspectives, especially in an area where we are very familiar with "accepted" thought. Indeed, it appears at times that the intention of this book is to take the reader as far away from accepted thought as is reasonably possible.

So, there are interesting viewpoints and perspectives, challenging concepts, and examples that offer a cultural viewpoint that the English speaker will find stimulating, and may find both thought-provoking, and inspiring. I particularly engaged with the sections on complexity, on the project as a complex adaptive system, and on *chrónos*-versus-*kairós* time. Some of these issues are starting to emerge as key elements in the future landscape of project-based management, and this text will assist in appreciating, understanding, and accepting the need for project management to be explored and documented in new and different ways.

Read this book. It will reward you in many ways, and will probably change the way that you think about project management in the future.

Dr. Steve Leybourne
Boston University

Acknowledgments

The contributing authors sincerely appreciate the individuals whose thoughts fostered and supported this book from conception to publishing.

We are grateful to Francesco Varanini and Walter Ginevri who have provided leadership to the project and promoted the very first edition of this work: *The Emerging Project Management: The Project as a Complex System*, Guerini & Associati, Milan, Italy, 2009. This English version of the book is nearly an identical translation of the original Italian edition.

We are thankful to Livio Paradiso for his dedication and hard work in leading the English translation. We express our profound gratitude to Diego Centanni for his vision and persistence in transforming our Italian publication into an international edition. Without the passion and timely insight of Senior Editor John Wyzalek and his team at Taylor & Francis Group, the English edition of this book would not have been possible.

About the Editors

Francesco Varanini spent three years as an anthropologist in Ecuador after obtaining his degree in sociology. On returning to Italy, he took a position with Mondadori, Italy's leading publisher, with responsibility for human resources development, organization services, and the full-text database of company products. Currently he works as a teacher and senior consultant in strategic and organizational development, new media, and social skills for change management. His projects include the Corporate University of Mobile Telecommunications Operator Vodafone and the Virtual Campus for e-Learning of Banca Intesa/Banca Commerciale Italiana (Italy's leading banking group), Internal Portal of Autogrill Group. He also is member of faculty, head of e-business and information and communication technology area, director of e-business management master, and director of customer relationship management master for ISTUD—Istituto di Studi Direzionali (Italian Business School). He is also managing editor of *Persone & Conoscenze* (http://www.este.it/res/rivista/rid/1/p/Persone+e+Conoscenze), adjunct professor in knowledge management, Università di Pisa, Corso di Laurea in Informatica Umanistica (digital humanities), and scientific director of Assoetica (www.assoetica.it).

Walter Ginevri, PMP®, PgMP®, has many years of experience as a senior consultant specializing in optimizing processes through the adoption of proven methodologies and tools. In this role, he collaborated with the most prestigious industrial groups and financial institutions providing advisory services within organizational turnaround programs.

He has participated in European research projects (Esprit, Eureka), contributed articles to several journals and attended, as an invited speaker, international congresses about project and portfolio management.

As founder and managing partner of PM for Complexity, he is a project advisor and executive coach for top firms in Italy, Germany, and France. He is also a senior trainer at the University of Verona and the Politecnico of Milan.

Genevri was appointed president of the PMI Northern Italy Chapter in January 2011. Since 2010, he has been a liaison member of the PMI Educational Foundation.

About the Contributors

Bruna Bergami, PMP®, works as a consultant, and as a project and a program manager with experience in mass retailing, IT, telecommunications, and energy. She is an expert in methodologies of software development, as well as commercial, quality and HR experience, acquired in different sectors (GDO, IT, telecommunications, and energy).

Gianluca Bocchi, professor of philosophy of science and epistemiology of human science at University of Bergamo, Italy, is the coauthor, with Mauro Ceruti, of several books among which is *La sfida della complessità (The Challenge of Complexity)* published in 1985 in Italy by Bruno Mondadori. He researches problems related to globalization, as well as the impact of new technologies on forms of knowing and communicating.

Diego Centanni is a Rotarian and project manager in the construction industry. He has lived and worked in Europe, Africa and North America. After his master's in engineering and project management, he learned to respond to a complex environment, in a perpetually unknown combination of circumstances and possibilities. He is now a project management specialist with more than five years of experience in project management, stakeholders' management, project costing and forecasting, development, and marketing. Centanni has a passion for challenges and is a lifelong scholar with an entrepreneurial spirit. Full details at www.dthere.com.

Luca Comello is a project office manager at Electrolux. He collaborates with Professor De Toni in research and publishing texts on complexity theories and the analysis of their implications in the organizational field. Among these titles are: *Prey or Spiders: Men and Organizations in the Cobweb of Complexity* (2005, UTET Università, Italy) and *Journey Into Complexity* (2010, Lulu).

Bice Dellarciprete, PMP®, PMI-RMP, has worked as a project and program manager for IT system integration and research projects. She also has worked on the planning, implementation, and start-up of enterprise project management processes and tools, and has focused on IT service management, studying the subject from a methodological point of view and working as change and release manager in a ISO20000-certified IT company. Currently, she works in a commercial role in the IT travel and transportation market.

Alberto Felice De Toni is a professor of operations management at the University of Udine, Italy, and is dean of the faculty of engineering. He is an author and coauthor of many publications including several books such as *International Operations Management: Lessons in Global Business* (2011, Gower Publishing), *Open Facility Management: A Successful Implementation in a Public Administration* (2009, IFMA—European Facility Management Network) and *Journey into Complexity* (2010, Lulu).

Fernando Giancotti is a major general of the Italian Air Force, and the defense attaché for the Italian Embassy in Berlin. He was formerly in charge of the staff policies, organization, and legal affairs of the Italian Air Forces. He published two essays in the United States and several articles and two books in Italy including *Leadership agile nella Complessità* (*Agile Leadership in Complex Environment*), published in 2008, by Guerini, Milan.

Mariù Moresco is a primary school teacher for many years and has researched various teaching methods and learning strategies. Over the past several years, she has applied project management techniques to education and trained a team of teachers for this initiative. Thanks to her cooperation with PMI Northern Italy Chapter experts, she designed and tested a methodological kit for project management in schools.

Stefano Morpurgo, an engineer, has worked for 25 years in management and project management, mostly with multinational companies. Since 2004 he has been a freelance trainer, coach, and project manager running IT and re-engineering projects for large public and private companies. He gained PMP certification in 2008, is an active member of PMI Northern Italy Chapter, and PMI-NIC director-at-large since 2010.

Carlo Notari, PMP®, has worked in several roles as both project and program manager in IT-related industries. Thanks to his participation in international initiatives, he acquired vast experience in organizational and management consulting. He was among the founders and inspirers of the PMI Northern Italy Chapter where he was president from 2003 to 2010, the year of his untimely death.

Livio Paradiso, PMP®, M.Sc., and an MBA student, has gained experience first in R&D projects and then in TLC services, processes, and projects. He is currently working as manager of project management for enterprise customers for the second largest Italian telecommunication services company.

Andrea Pinnola, PMP®, is a project manager who has been working for 15 years on several IT system integration and research projects. Passionate about quality and people development, in 2000 he received the Human Resources Development Award from his company for results in the development of his team. He gained PMP certification in 2008.

Michela Ruffa, PM senior consultant, has been working for 20 years on the definition, implementation, and start-up of enterprise project and portfolio management models, processes, and tools in complex industrial and ITC organizations. Her responsibilities often include training, coaching, and change management. She gained PMP certification in 2005 and has been an active member of the PMI Northern Italy Chapter since 2003.

Roberto Villa's, PMP®, liberal arts education, technological experience and organization mindset are what distinguish Roberto. He has years of experience in both IT innovation projects and implementation of Lean methodologies in telco contexts. His curiosity led him to explore worlds, through small companies and large corporations. He then moved to project management, in a chaotic but not random way drawn by an invisible thread connecting passion, ability, and opportunity. He applies what he learned—and believes in—through his company, Projectize, with work, training, associations, conferences, testimonies, and writings.

Introduction

This book is about projects and complexity but, because of its birth, it's a project in itself as well: an initiative aimed at reaching a goal within limits of time and cost. It's also a complex system: it has been drawn up with the contribution of several authors, who started comparing each other's experiences, and at the same time definitely influenced each other. Its origin lies in different processes as well, such as individual writing, dialogues, and teamwork.

Everything started from a proposal Francesco Varanini made to the PMI® Northern Italy Chapter (NIC), a professional association he had been co-operating with for a long time. His idea was to study the nature of a project as a complex system.

The approach to projects is usually well established. The method used, somehow limiting, but functional and reassuring, requires a previous description of all the activities deemed essential to reach the goal, first of all, then their even distribution over a certain period of time. But according to project managers' experiences, simplifying the process like this doesn't always add to effectiveness. Actually, seen under another light, a project appears as an organizational network, looking different according to the observer's point of view; a project takes a definite shape only in the course of time. Further on in a project, the whole picture doesn't simply amount to the sum of its parts, and managing a project means moving to the edge of chaos by trial and error. Therefore, inasmuch as we had started considering and looking at a project as a complex system, our challenge was to find in studies about those kinds of systems, attitudes and instruments that could be profitably used by project managers.

The PMI NIC Association's answer was prompt; the research very soon became an interdisciplinary one, assigned to a team composed of NIC members under Walter Ginevri's coordination. Calling themselves "Complexnauts" (another example of complexity might be that there are 12 of them, just like Jesus' Apostles), the NIC project managers— Bruna Bergami, Diego Centanni, Bice Dellarciprete, Walter Ginevri, Mariù Moresco, Stefano Morpurgo, Carlo Notari, Livio Paradiso, Andrea Pinnola, Michela Ruffa, and Roberto Villa—started their journey in the second half of 2008, guided by Francesco Varanini as mentor.

On their way, marked by periodical meetings and supported by a Web Environment 2.0 used to exchange and share the knowledge thus far acquired, the "Complexnauts" made lucky encounters, first of all with four expert travelers—Gianluca Bocchi, Alberto Felice De Toni, Luca Comello, and Fernando Giancotti—who in completely different domains, philosophy of science, new models of company management, and new models of military leadership, had much earlier already accepted the challenge of complexity. The conference "Projects and Complexity," held in Milan in November 2008, represented an important milestone on the route, and at the same time was the opportunity to present the work to the public. This event was the first one dedicated to this subject in Italy and its success encouraged NIC and its chairman to carry on with even more energy.

The project of this book was born at that point. The book represents a complex object in itself, readable in different ways: its first part is an essay, exploring the philosophy and culture of projects from various points of view. The second part is a working tool, a toolbox, almost a handbook intended for project managers. This book primarily addresses professional project managers, but offers remarks and stimuli to all interested in corporate management and functioning. In fact, project management, as the management of mutable complex organizational systems, aiming nevertheless at one specific goal, appears as the paradigm of a new way of managing any enterprise.

As with all complex systems, this book is now completing the preparation stage and entering its real life. We expect it to evolve thanks to its creators' and also its readers' contributions and to all the people who will feel like taking part in this fascinating journey. To know more about our journey, please visit http://www.projectsandcomplexity.org

1

Complexity in Projects: A Humanistic View

Francesco Varanini

CONTENTS

PROJECT

The Latin prefix *pro-*, "forward," points to an Indo-European conception to which we owe the front, or "prow" of the boat as well: *progress*: "to walk forward"; *process*: the steps of those who *pro* (forward) and *cede* (ceed, give way without opposing any resistance); *produce*: "to bring forward," "to bring out"; *program*: "written beforehand"; *planning*: "designing on a plane surface"; *prophecy*, too: "foretelling."

The project is part of the same context: from the Latin *pro-jacere*, "to throw forward, to project," to throw an arrow or a spear, in the way we throw dice. This expression reaches every language, even faraway ones: in Japanese it is called *purojekut*, through the French language. *Projecter* is found at the beginning of the fifteenth century. In the middle of the sixteenth century *project* replaces the former *pourget*.

A few years later, in 1553, in the *Giornale dell'assedio di Montalcino* (*Journal of the Montalcino Siege*) we find the word used in Italian for the first time. The meaning is not strictly the current one yet; *progetto* standing for "indefinite or peculiar plan, hardly feasible." Only at the end of the sixteenth century does *to plan* become "devising a plan and find ways of carrying it out." But at last in the nineteenth century the verb speaks exactly of "designing a building, making calculations and plans for its execution." And yet (at the end of the twentieth century) purists of the Italian language oppose the expression. For once they are not completely wrong: unlike all other languages, the Italian words *progettare* and *progetto* only partially convey the Latin and French meanings.

The English *project*, the German *projekt*, the Russian *proekt*, the Spanish *proyecto*, the Portuguese *projeto*: in all languages this expression still conveys the sense of "casting" something, therefore that of giving out rays of light or "reproducing images on a screen." A single word, a single verb, expresses at the same time a military, geometric, psychological, and cinematographic idea. In Italian, on the contrary, the idea of *project* has been divided in two: on one side we have *proiettare, proiezione* (to project, projection) and on the other side *progettare, progetto* (to plan, project). Therefore—even though it is wise, when possible, to express ideas in a maternal language, not a merely technical one, a language making sense in everyday life too—when one wants to refer to "project" in the proper full sense, the use of the English word *project* is better: because I am not just planning, as we would say in Italian. I am also projecting myself into the future into a place that never existed before. I'm casting a projectile toward a target I cannot see.

The project, devoid of the idea of projection, missing its analogy with the projectile, removed from the image of the spear and the arrow, deprived of the idea of a light cone cast on a dark place, of the heart put into every effort, could appear to have lost its meaning, lacking emotion, a merely technical, abstract, faraway matter. Not so useful, and ineffective.

PROJECT SQUEEZED ON A PLANE

The French language, in the sixteenth century, starts using *plan*, from the Latin *planum*, Indo-European root *pela-* (flat, outstretched), with the meaning of "something represented on a flat surface." Hence in the seventeenth century, the Italian and English words: in this language, the verb (planning) gains ground.

But the sociological and economic meaning goes back to much closer times: The Twenties of the past century: the economic cycles of the *laissez-faire* market deteriorate in an out-of control crisis, and the answer is sought through economic politics and target-oriented development projects. Thus the Soviet Union, in 1921 sets up the *Gosplan, Gosudarstvennaja Planovaja Komissija* (State Planning Commission), and in 1929 the first Five-Year Plan is launched. In the same years, Western economies (as in Roosevelt's New Deal) look for a new agreement between social partners: this agreement will have to become, here as well, a plan based on the persuasion (or conceit) of being able to direct the economy toward pre-established goals.

The *Economist*, in its March 30, 1935 issue, indeed writes that if private companies clearly failed in taking the right steps, then "planning" must be tried, a new expression the journalist writes, with reason, in quotes.

It is surely useful, and maybe essential, to write programs (in Greek *progràphein*, "writing before") or documents when the decision as to how to face the future is taken, in which one explains—to those who made demands, to those who will have to bear the expenses, in general to all those concerned—what will be done.

Planning, a schematic and brief but also not exhaustive representation of the future, is of great use. Indeed, as soon as an image of the future is put down in writing, one must remember that the future is still unknown, and that our possibilities of preordaining it are limited; our ability of directing development one way or another depends on variables totally beyond our possibility of control.

On the contrary, sometimes we prefer not to see. Fear of future negative events is fought by giving a preliminary definition of what the future will have to be. So the future is not "what will be" but what I decide today will be tomorrow. We don't accept, in doing so, the future for its novelty: it won't be possible to accept unexpected aspects of the future as blessings, as riches, because the future will be under control, that is, compared with the forecasting I am doing today.

In other words, using planning in a defensive way, the project is debased; the project, squeezed on the plane surface, considered as a predetermination of the future, is liberated from uncertainty. But it is also deprived of essential aspects: hope, dreams, and novelty.

PROJECT SUBMITTED TO CONTROL

And afterwards, the reduction of the project in the plan opens the way to another, even more dangerous reduction: *control*.

In France, in the thirteenth century, the *rôle* is the "register": the list of the members of an organization, the description of some proceedings. How to check that the contents of the register exactly reproduce reality? Making another registration, keeping a counterregister, a *contre-rôle*. The verb *contreroller* is then coined (as in French, in Middle-Ages Latin: *contrarotulus, contrarotulare*). Hence in English, as early as the fifteenth century, *control*.

Control is certainly a useful activity. Checking registrations, making sure that the number of subordinates, the costs, and the revenues have been correctly calculated, is dutiful and necessary. As long as the limits of control are not forgotten: the behavior under the close examination of control is not the best behavior that is possible. It is only the behavior conforming to the rule. This is why sometimes orientation to control rewards the worst management: it rewards those who, hiding behind conformity to the rule, avoid assuming any responsibility and do not catch hold of opportunities for improvement.

We should also remember that the rule (translation of the Latin *norma*) is useful, even required, in an organizational, administrative, and accounting context. It's more difficult, more dangerous, to consider the rule, therefore the "normal behavior," as required inside a project. *Norma* (hence "norm") is Latin for "squadron, team." *Exquadrare* means "to reduce to a square shape." In this case, to the implicit compression on the plane, the representation through a one-dimensional image, a further compression is added: the shape the object I have in mind might take doesn't matter; the only shape I consider is the square one.

No one denies the necessity of "squaring accounts." Possibly the regular management of a company may be subjected to a single shape, maybe oriented to squaring. But if we have in mind the project, something complex and multiform, its one-dimensional, one-linear representation can but appear reductive.

Submitting a project to control means comparing what we are doing with a program, a document "written before." In this way of considering a project, the attention to the construction part, to the invention, to the discovery of something that wasn't there before, is limited to the starting phase. The "time for the project," the time we grant ourselves to look at the unknown, in which we afford to imagine what isn't there, is over very soon. The written descriptions of the "things to do" end it.

From this point on, following what has been written and signed, one executes and carries out what is "prescribed." One expects all activities to consist of what is described in a detailed comprehensive picture drawn beforehand, and only in that. One expects activities to be carried out in that sequence, keeping to the calendar defined earlier. Any deviation from the course previously outlined is considered a flaw; any variance is something that will require a justification.

One thinks again, goes back to the initial thought and modifies it, only if it's essential. As if thinking and working were two activities not to be carried out together, as if, in order to start carrying out the work required to bring the project to a positive conclusion, thinking should be stopped. A project so reductively intended is born with the hand brake on. Tied to the starting point, the project will be turned backward instead of looking ahead.

PROJECT MANAGER'S ROLE

When a project is conceived in the light of planning control, when a project is divided into an initial phase of planning and a second phase of realization, the project manager is first of all considered the person responsible for its execution. The project manager takes in his hands a project already defined as to its preliminary scope (which is its target) a project already constrained (as to predetermined time and cost), and engenders, from a practical and material point of view, its execution.

I nevertheless believe experience and self-awareness, an awareness of the actual substance of work he's carrying out, will help the project manager realize his work isn't restricted to his overseeing its execution. Because, more than anyone else, the project manager is aware that the muddle can be clarified in several ways. He knows that "the way of doing things" finally adopted almost never coincides with what had been seen beforehand. He knows that the definition of aims and the execution cannot actually be conducted separately. He knows a plan always contains many imperfections and that only later will the way of putting them right be discovered. He knows no one can possibly declare when and how a project can reasonably be considered completed.

For this reason, I think, any project manager feels an uncomfortable loss of contact with reality when he's compelled to devote himself to the compilation of documents containing detailed estimates of what is unpredictable, when he has to write down detailed calendars of what will be done every day for years to come, and when he has to draw up on paper an accurate comparison between what was foreseen and what has been done, and when he has to state the exact measurement of how much of a project has already been executed.

The attention to planning and control is only right: nevertheless, treading this road means deviating from the project. Here is where, I believe, the annoyance and frustration originate: so much time, so much attention devoted to cultivating a fetishist image of the project. The center of attention is not the project anymore. In the place of the project, entangled and partially obscure, yet alive, there is a simplified image, a fetish removed from the projection, from the eyes looking ahead, from hope, from the dream of what we undertook to realize; a fetish removed from what customers and all stakeholders involved in the project expected; a fetish removed from the actual project, the one lived through day by day by those who work on it. One ends up by worshipping an image, by working around a project that isn't there; one ends up by moving inside an illusory reality.

Annoyance and frustration are born from this awareness: how much wasted energy! What a great amount of time used simply to square things on paper. What a lot of attention diverted from the project in its full reality. And all this is because one is held prisoner by a linear idea of a project, because to face its complexity one looks for simplification, and because all available tools are directed to the definition of the goal, once and for all: to "secure" it. And what's more, they are control-oriented.

FEAR AND SELF-DECEIT

This behavior, I think, depends on our being faced, when considering the very idea of a project, with a complex situation, with a set heavy with uncertainties and risks, with a muddle of thoughts and actions that can only generate anxiety and fear.

Faced with this muddle, we choose to substitute the reality of things (we actually don't know what will happen and how to act in order to reach the goal) with a simplified image, a one-dimensional image, flattened on a plane. And on that plane, is a squared shape, clean-cut, well-trimmed, and reassuring. So in our eyes the muddle becomes a thread and links the idea we set out with to a specific result.

It's a pity this is an illusion that we use for self-deceit. We all do, all of us working on the project, expecting the project to answer our needs, meeting the costs of the project, all of us somehow involved in the project, because we are aware that the muddle can be clarified in many different ways. We are well aware that ours is but one of those possible ways, because we are very well aware there's no ideal way of clarifying the muddle, one that can be defined "better than the others." We know there are different ways, each bearing advantages and disadvantages. We know very well we cannot say beforehand exactly what outcome will be reached, how much the costs will be, or how long reaching the outcome will take. The project constrained in a shape beforehand is not a project anymore, because the project is in itself denial of the rule. The project means going beyond.

COMPLEXITY

If I look at a project trying to set aside the overly simple picture I have in mind, I don't see an effective system, working by clear logic, where the whole equals the sum of its parts. I see a living organism, a defective one, in evolution, unceasingly metamorphosing. I see a muddle.

It makes me think of *Il Pasticciaccio* (The Awful Mess) by Gadda [1, p. 5]. Right at the beginning of this novel, Officer Ingravallo's investigation method is explained:

> He sustained, among other things, that unforeseen catastrophes are never the consequence or the effect, if you prefer, of a single motive, of a cause singular: but they are rather like a whirlpool, a cyclonic point of depression in the consciousness of the world, towards which a whole multitude of converging causes have contributed. He also used words like knot or tangle, or muddle, or *gnommero* which in Roman dialect means skein.

Calvino in *Six Memos for the Next Millennium* [2, p. 106] talks about it: Gadda "sees the world as a 'system of systems,' where every single system influences the others and is influenced by them." Gadda describes the world "without any moderation of its inextricable complexity, or rather the simultaneous presence of the most assorted elements contributing to determine each event."

It's hard to find a more effective description of project: the project is a system of systems where each system influences the others, and is at the same time influenced by them; it's a knot or muddle appearing inextricably complex to us. The solution—just one of the possible solutions—appears mysterious. Only if we accept seeing the muddle, if we accept its inextricable complexity, only if we are able to live with the anxiety and the powerlessness connected to "not knowing how to do it," will we be able to build effective knowledge and obtain a good result.

I'm deliberately speaking of result. If I accept the project in its complexity, I know that at the moment I can't see any point of arrival. I have no way of knowing how and where the project will "end up." Having specified the goal—literally, the target—means having expressed the intention, having imagined a possible future. But even I, as the other people involved do, will be able to see in different moments of the project life different pictures, all coherently consistent with the departure point, all similarly coherent with the goal. If I keep to the picture of the future I can see in a particular moment, and then I derive from this single picture the work intended for the goal, I limit myself without reason. Accepting complexity is much better. That is, accepting that while I work the goal appears more distinctly only sometimes, The goal is not "what it was supposed to be"; it's the result that moment after moment I seem to be possibly reaching.

Therefore "the result is implicit in the process." The most effective way of reaching the goal is accepting the result (Latin *re saltare*). The result is what stands out, literally what "bounces back, turns out," and binds the situation under my eyes to the point of departure and to the possible point of arrival. We look forward to the result. To look forward to is "to turn to with our heart," or "to incline toward."

For these reasons, looking at the project, we can speak of exactitude. To exact is to act outwards (in Latin *ex agere*) "to push outward." The result I'm inclined to is pushed outwards, moment after moment, by my work. The result is an *emergence*. To emerge is the Latin word for "to come out, or to come out of the water." Legal language extends the figurative use: "*si aliquid inopinatum emergat*," in case of something unforeseen, something nobody thought of before, emerges (comes afloat).

PROJECT MANAGEMENT BEYOND MANAGEMENT

There is already enough for us to understand how the optics of complexity changes our view of the project manager's role. The project manager enjoys high technical consideration but still appears somehow subordinate, compared with the figure of the fully recognized manager. On the contrary, not only do we consider the project manager first of all a manager, as all others, a manager in every respect, but in our legitimate opinion the project manager is the manager of the future, the figure allowing us to look beyond the limits of management, beyond that management, which, in its inadequacy, impairs the company.

> Although the manager may, at least somehow, set the goals for himself, the project manager has to face preset goals. The manager may modify the regime of constraints, but the weight of the constraint to be complied with is taken on by the project manager.
>
> Although the manager may follow the wind, hide in the crowd, justify himself saying he followed some consultant's indications, the project manager has no way out. The task, and the engagement undertaken, are clear enough.
>
> Although for a manager the alibi of unfavorable and unforeseen outer conditions may work, for the project manager there's no possible alibi; the goal must be reached anyway.

So, by observation of how the project manager works day after day, the management of the future can be outlined, and the management needed to build the future can be described. All this is true, however, if, and only if, project managers, instead of conforming to a simplified and schematic image of the world, tied to the past, accept responsibility for the complexity of the project.

This is how, apart from any display of management, the project manager's role is revealed as care, guidance, and governance.

> Care: The attention paid to taking care of others and procuring what they need
>
> Guidance: Showing "how it is done," not as a boss but walking alongside as *primus inter pares*, first among equals
>
> Governance: The gesture of the steersman who knows where the port of destination is and "stays the course" by subsequent adjustments

PROJECT MANAGER AS A STORYTELLER

What has been said so far, has been said by an observer. Any observer influences the object of his research and is subjected to its influence, in turn. Each stakeholder carries his own point of view. From that point of view, any actor observes the project. We could say that every actor illuminates with the cone of light of his glance a common territory, the project. All points of view are valid. All looks contribute to the illumination of the scene. Every look expresses itself in the shape of a story. Everyone has the right or duty to contribute to the project with his own story. All stories contribute to knowledge.

The Indo-European root *gn-/gen-/gne-/gno-/* speaks about "noticing," "learning with the intellect," "knowing something," and therefore "knowing." Hence the Greek *gignoskein*, "to know," *gnosis*, "knowledge." In Latin we have *co-gno-sco* (where *co-* means "with," *-sco* stands for "starting to"), *gnarus* (knowledgeable, expert), *notio*, and *notitia* (knowledge). In ancient High German from the root *cnaen*, *-cnahen* only compound verbs are derived, but this is the way *können* (can, to understand), and *kennen* (to know) are derived in Modern German. In ancient English you find *gecnawan*, then *cnawan* (hence to know) but also the auxiliary *can* (have power to, be able). In any case, we're not talking about codified normalized knowledge described by specific models. From the same Indo-European root (through *gnarus*), we find also *(g)narrare* (to narrate).

The project cannot be totally dominated, described from above, from outside. It can only be told as a story. The project manager, as any other stakeholder involved in the project, bears one point of view, and from that point of view acts as storyteller. But, furthermore, by his experience and aptitude, on account of his special ability in reading weak signals, in observing the unknown, he has been entrusted with a task distinguishing him from the stakeholders: he has to hold the cloth that is being woven together. He is the rhapsode, he who "weaves the song." Therefore, from now on, to tell how we can comprehend and live the project in the light of what the considerations about complexity teach us, I continue by telling stories.

BEST OF POSSIBLE PROJECTS: LEANING TOWER OF PISA

In 1063, when the town was outstanding in the Christian world as to military and political power, as to economic prosperity, and cultural production, the inhabitants of Pisa decided to start building the new cathedral. Inscriptions embedded in the cathedral façade praise the architect Buscheto, buried in a recarved Roman sarcophagus that is part of the façade itself. Buscheto is celebrated as greater than Daedalus or Ulysses, not only for his pure art, but also for his attitude to the project: the ability with which he took care of the transportation by land and sea of enormous columns, avoiding at the same time Genoese and Saracen ambushes, is especially praised.

Buscheto's praise reaches its height in the verse: *"Non habet exemplum niveo de marmore templum,"* Latin for "this white marble temple has no prototype nor comparison." The discussion, here, is about the exceptional nature of the work, where a specific style, Romanesque Pisan, is put into practice in order to emulate the great architectures in Imperial Rome and Islam, and competes, as to proportions and complexity, with Christianity's greatest temples, the first St. Peter's Basilica, and the Hagia Sofia in Constantinople. But what is discussed is also an essential aspect of any project: *"non habet exemplum."* Every project is an unique work, similar only to itself, a prototype, a masterpiece.

In 1118 the cathedral, the Duomo, is consecrated. In 1152 work for the Baptistery begins. Here as well there are eventful journeys by sea; it's hard to carry the big monolithic granite columns used for the building. The way the stakeholders, who are, after all, all the inhabitants of Pisa, get involved, holds our attention. So, the chronicle goes, in 1163 a monthly tax of one denarius per family is instituted, and compulsory shifts are organized in each district for the setting up of the columns. On August 9, 1173, the foundation stone, the *primarium lapidem*, of the bell tower is finally laid.

Ten years later, when the building gets to about one and a half meters from the third terrace, the work must be suspended. The bell tower, due to the unsteady and marshy ground, is now frighteningly leaning on one side, despite the adjustments of the slope enforced during the work. A sad solution is chosen: the broken leaning tower is covered with a small dome, and the bells are set up there.

Fifty years will go by before, having verified that the stump is solid and the sinking is not worsening, the possibility of opening the building site again is considered. Work is resumed, probably between 1272 and 1275. But the inclination increases, although irregularly. So, again after about 10 years' work, now that the fifth terrace is reached, and in spite of the care in counterbalancing the inclination with static choices that best exploit the available technologies; a makeshift solution must be chosen again. At that level a place for the bells is found and the work is suspended.

Legend, rather than history, has it that this occurred in the most ill-omened year in Pisa's history. On August 6, 1284 the Republic of Pisa was defeated by the rival Republic of Genoa in the battle of Meloria (the low rocks off Leghorn). That didn't mean only the loss of 49 galleys, that didn't mean only the 6,000 dead, the over 10,000 prisoners, for Pisa that meant the beginning of its decline.

Nevertheless the work on Piazza del Duomo goes on. In 1277 the foundations of the last building of the wonderful square, the Camposanto (the graveyard), are laid. But for a new opening of the tower building site the wait is longer. The sources, full of gaps, date the last resumption around 1370, exactly 200 years after the beginning of the work, and 100 years after the second cycle.

A sixth loggia was added to the five terraces already built, and over the sixth loggia, with the plain function of the final element of the building, the belfry. The adjustment of the slope, already visible in the sixth terrace, is clearly perceptible in the belfry.

We know the names of the first designers and builders of the other three buildings: the Duomo, the Baptistery, the Graveyard; but not for the Leaning Tower. Neither an inscription on the building, nor a document by the local historians gives us a precise indication about the name. We reasonably have to ascribe this silence to shame or to censorship, still as a consequence of what was interpreted as a serious technical accident: the tower leaning on one side, the project that, in its realization, fatally deviates from the project planned beforehand.

In 1800 works were carried out and part of the buried basement was brought back out. This was a serious mistake, the stability of the tower changed and the leaning process quickened. Again, between 1990 and 2001, a new building site was opened. Once more, the most advanced technologies of the time have been adopted to keep the building standing, in spite of the slope. The lateral shift of the axis, that exceeded five meters, is now reduced to less than four. The building site is not expected to be opened again for 300 years. But it's better not to go too far with forecasts because the Leaning Tower is in itself a challenge to forecasts and to the project purity. The slope itself, that is, the variance from the project as it should have been carried out, proves, beyond any pretense of control by the planners, to be the origin of the good quality of the work; the source of its unexpected, or rather, unpredictable, beauty.

In fact, in the 1830s, a strong dispute took place. Because, according to a well-disseminated concept, there can be no good result that is not included in the project, a scholar doggedly maintained that the slope of the tower had been designed, and was calculated in the project. He was soon belied [3].

LEANING TOWER LESSON

The project can be put into practice only by accepting what the environment suggests or forces us to do. This doesn't mean one has to give up forecasting: setting a target is always useful and constructive, expressing expectations and setting them as time, place, cost, and mode (or technology) constraints is always useful. But one has to be constantly available to the revision of any aspect of the forecasting. Only by giving up attachment to what had been designed at the beginning is an outcome allowed to see the light of day. The best possible project is a project being transformed into something that can exist in the world, something that can find a place not in an ideal world, but at that site, at that time. This is how the Leaning Tower, the best of possible towers, exists in the twelfth century, how the thirteenth century tower exists, the nineteenth century tower, the twentieth century tower, and the twenty-first century tower.

The story of the Leaning Tower speaks of environment: the collapse of the ground depends on its nature in that place. It's not worth speaking, even in a precise technical way, of soft normally consolidated clay: this would already be a generalization and a removal from the environment. I have to build the tower in that place, at that time; only that tower can be the Leaning Tower.

The story of the Leaning Tower speaks of technology; the technology to build the tower in that place, at the moment the work started, has been exploited to the limit. But that was not enough to complete the project. The 100-years-later technology, or the 200-years-later one, allowed things to be seen differently. So did the end-of-the-millennium technology; every time the Tower was seen in a different way, and every time solutions previously unthinkable were designed.

The story of the Leaning Tower speaks of goal: the goal can only be reached with approximation. Goal literally means "target," thinking, therefore, how the very idea of target can be read in a different way at a close distance can be enough. There are fixed and moving targets, visible and invisible targets, and also targets that are out of reach, as they are protected by a covering mass. And if we look at the target of target-shooting, too, we see a system of concentric circles and sections. I can hit the target at different points and I still make the hit. There's no way I can hit the center of the target perfectly. And there's no visible reason to waste my resources in order to try to get closer to this anyway unreachable center.

Such is the Leaning Tower, as we can see it: the Tower has been built, carried out, and the target has been reached, although at the price of renouncing an aspect that the project on paper considered unquestionable: its verticality. But this limit appears now as an added value. The story of the Leaning Tower speaks of hindsight: only with hindsight, resulting from different judgments and readings, from the interpretation of different people, taking place in different places, times, and cultural contexts, unforeseeable beforehand, and even from the field the project managers can manage, can the project be understood.

The project starts being revealed, starts revealing its secrets when the work on paper stops. The project is revealed only in the course of its being carried out. The project is clearly outlined only afterwards. Only afterwards can we be aware of what we did and only afterwards can we tell the story, speaking of the course we took. The story of the present cannot be told; the present can only be lived through. Only when the planning work is over and the project becomes something, can this something be used. Only with use does one understand the essence of the thing itself.

The aesthetics of the project is something that emerges, as well: as victims of a certain idea of a project, beauty for us becomes compliance to the forecast, and, faced with the incontrovertible beauty of the tower, we are tempted to maintain that the project foresaw a leaning tower. On the contrary, the tower, as it can now be seen, was born out of the existence of an alternative: just to get close enough to the goal, deviations from the plan had to be accepted, deviations leading to frustration, disappointment, and guilt. Nevertheless, what seemed in the moments when one couldn't help adapting the project to the situation, a renunciation, a flaw, and a limit, to our eyes now appears, compared to any other tower, as the Leaning Tower's "competitive advantage."

MOTHER OF ALL PROJECTS: TOWER OF BABEL ACCORDING TO DANTE

Any reasoning about the intrinsic complexity of a project cannot help taking into consideration an image, a literary *topos*: the Tower of Babel. No project is more ambitious. Man challenges his own God in order to build "a city and a tower whose top may reach unto heaven" (Genesis, 11:4).

To the Lord this intention appears as an unacceptable gesture of human arrogance; in his eyes man shows here to have lost the sense of limits. Therefore He confounds the languages "so that they may not understand one another's speech" (Genesis 11:7). Thus the project fails: men, forced by the impossibility of communicating, "left off to build the city" (Genesis 11: 8). Dante, in his *De vulgari eloquentia* [4] speaking of the vulgar language as the (possible) perfect language, cannot avoid referring to the passage in the Bible. After him, Bacon and Comenius and Descartes and Leibnitz and Humboldt, up to Umberto Eco, don't fail to draw on this passage when they discuss languages and dialects and the possible origin of all languages from a single stock.

But Bacon and Comenius and Descartes and Leibnitz and Humboldt, up to Umberto Eco, all give an interpretation of the passage in the Bible thinking of natural languages, mother tongues: each person can only fully express himself in his own language; all languages are different, and they have in mind the attempts at creating an interlanguage, a language allowing universal talk.

Only Dante intends words as part of the working activity; he looks at language as an instrument to express professional knowledge. According to him, the *confusio linguarum* (confounding of languages) is an organizational problem, a limit invalidating the project.

> *Presumpsit ergo in corde suo incurabilis homo (...) arte sua non solum superare naturam, set etiam ipsum naturantem, quid Deus est, et cepit edificare turrim in Sennaar, que postea dicta est Babel, hoc est "confusio," per quam celum ascendere, intendens inscius non equare, sed suum superare Factorem. (...) Siquidem pene totum humanum genus ad opus iniquitatis coierat: pars imperabant, pars architectabantur, pars muros moliebantur, pars amussibus regulabant, pars trullis linebant, pars scindere rupes, pars mari, pars terra vehere intendebant, partesque diverse diversis aliis operibus indulgebant; cum celitus tanta confusione percussi sunt ut, qui omnes una eademque loquela deserviebant ad opus, ab opere multis diversificati loquelis desinerent et nunquam ad idem commertium convenirent. Solis etenim in uno convenientibus actu eadem loquela remansit: puta cunctis architectoribus una, cunctis saxa volventibus una, cunctis ea parantibus una; et sic de singulis operantibus accidit. Quot quot autem exercitii varietates tendebant ad opus, tot tot ydiomatibus tunc genus humanum disiungitur; et quanto excellentius exercebant, tanto rudius nunc barbariusque locuntur.*
> *(De vulgari eloquentia, I, VII, p. 14, of the English translation)*

[Incorrigible humanity, therefore, presumed in its heart to outdo in skill not only nature but the source of its own nature, who is God; and began to build a tower in Sennaar, which afterwards was called Babel, that is, "confu-sion." By this means human beings hoped to climb up to heaven, intending in their foolishness not to equal but to excel their creator. (…) Almost the whole of the human race had collaborated in this work of evil. Some gave orders, some drew up designs; some built walls, some measured them with plumb-lines, some smeared mortar on them with trowels; some were intent on breaking stones, some on carrying them by sea, some by land; and other groups still were engaged in other activities—until they were all struck by a great blow from heaven. Previously all of them had spoken one and the same language while carrying out their tasks; but now they were forced to leave off their labors, never to return to the same occupation, because they had been split up into groups speaking different languages. Only among those who were engaged in a particular activity did their language remain unchanged; so, for instance, there was one for all the architects, one for all the carriers of stones, one for all the stone-breakers, and so on for all the different operations. As many as were the types of work involved in the enterprise, so many were the languages by which the human race was frag-mented; and the more skill required for the type of work, the more rud-imentary and barbaric the language they now spoke.]

The passage in Genesis 11:4, "let us build a city and a tower whose top reaches unto heaven," is especially vivid in the narration and particular care is given to the details, in a way that should in itself make us think: "Some were giving orders, some were planning, some were building walls, some were squaring them with levels, some were plastering with trowels, some were breaking stones, some were taking them through land and sea; different groups were intent in different works."

Dante, in short, doesn't speak about languages, but about professional families. And he makes us wonder: how can these different activities be held together, how can they be read as a whole, as a process aimed at a single project. Dante shows us in few words the extreme complexity of the work in which these ambitious men are engaged. Dante tells us, as well, that in order to build the city and erect the tower, each single profession will have to be exploited to its utmost.

Now, every profession manifests itself when it best processes its know-how, and the elaboration, which happens inside, expresses itself in the culture of the work, in specificity, in a way of thinking, and in pride, too: attitudes justified by the very professionalism, making sense only inside the group of those practicing that profession. Taking the trouble of being understood by others is, from the point of view of professional growth and development, a waste of time and resources.

To a careful observer, all this is an enormous and fatal paradox. The more one gets into his world the better he works. But in doing so one is removed from others, the stone-cutter is a good stone-cutter if he feels different from the bricklayer and doesn't waste time in making comparisons with the latter, and the other way round. But then the stone-cutters' and the bricklayers' work will be efficient if seen as part of one system. The more complex is the work, the more this is needed and becomes at the same time more difficult.

So, Dante appears to be saying, the tower builders fell into confusion:

> They were struck with such confusion from heaven that, while they were all devoted to their undertaking when they were using the same language, they became different because of the many languages, they abandoned their work and could never again retrieve the same mutual understanding, that ability of working together. Only among those who were engaged in a particular activity did their language remain unchanged; so, for instance, there was one for all the architects, one for all the carriers of stones, one for all the stone-breakers, and so on for all the different operations. As many as were the types of work involved in the enterprise, so many were the languages by which the human race was fragmented; and the more skill required for the type of work, the more rudimentary and barbaric the language they now spoke.

Here stands the catastrophe, the disgrace. Professions continue to develop, to become more articulate on the inside. But, out of envy, or out of contempt, or because it seems impossible or because it appears useless, a dialogue or common goal is missing. Everyone is speaking for himself, sees only his own sectional project, and the whole project only from his own point of view. "*Et quanto excellentius exercebant, tanto rudius nunc barbarisque locuntur*" ("And the higher was their practice of their activity, the rougher and more barbarian appeared their way of speaking").

It's a perverse spiral: the deeper and richer and more specific are the contents of my professionalism, the more they will appear incomprehensible and closed to those who do not belong to it; and the more a profession is self-aware, the more those who do not share its values and don't belong to it will seem barbarians. There's a wide cultural distance between architect and stone-cutter and manual laborers and bricklayers: they all think differently and therefore express themselves differently. They will never be able to communicate if not through an interlanguage, a vehicular auxiliary language.

And still, if "a common understanding," a shared language is missing, dialogue is missing; if interaction between different professional families is lacking, the project will be like Babel, the erection of the tower will become impossible and will have to be abandoned.

PROJECT MANAGER AS A SUBSTITUTE FOR THE ABSENT GOD AND THE PROJECT AS TOWER, OR: PRAISE OF IMPERFECTION

In a faraway time, we can read in the Bible, "[E]veryone was using the same language and the same words" (Genesis, 11:1). At that time maybe good communication and unity of intent could be taken for granted. But after God condemned us to the *confusio linguarum* nothing could be taken for granted anymore. God compelled us to work in an uncertain world, totally out of our hands. We are condemned to complexity.

The project deserves the best technicians to work on it. The best technicians, however, are conditioned by their pride; this compels them to express themselves and behave in such a way that the difference is emphasized. The sum of differences creates bedlam: the impossibility of communicating and of putting together the pieces of work everyone completes singularly. Shared work seems impossible; the project fails.

We can, however, observe that every project possesses its own popular, common, even if imperfect, language, offering all the people working on the task the possibility of sharing knowledge and experience. Without this idiom, this tool for mutual understanding that joins those working on the project, the awareness of a unity of purpose cannot be constructed. But this idiom comes to light, gains ground, and thrives only if someone takes care of it. And this heralds the importance of a project manager aware of how complexity can be reduced if, although with respect to each professional family's independence, the dialogue between professional families is encouraged.

At this point an extremely wide and essential field of action opens for the project manager: he is endowed with the role of substitute for the absent God. His concern is to guarantee, notwithstanding the symmetric pride of the different professional families, dialogue and co-operation. As it seems, God had no intention of denying professional pride some space, but he wanted to reduce it, removing harshness, excesses, and sources of conflict.

Pride is a treasure, but one had better keep within the limit, because any excess diverts from the tension toward the outcome. Figuring oneself as God may be constructive: without daring, the project remains prosaic. But we should be aware of the limit, a limit that is particularly difficult to sense, but still, is there. Because no one among us is God. The Tower of Babel tells us this, too. In Hebrew *Bābhél* literally means "God's door or gate."

The Tower is a traditional Mesopotamian building, a *ziggurat*; built in bricks and bitumen, it rises in terraces, it evokes the shape of a mountain, and it is a bridge between earth and heaven. The *atanor*, the oven used by alchemists for their transmutation experiments, was not shaped as a ziggurat by accident: the passage from lead to gold brings back to mind the passage from earth to heaven; in both cases it means going beyond limits, *superare naturam* (Latin: to overcome nature). In both cases, it's a matter of projects. Every project, taken in its complexity, is transmutation, alchemy: the ability of challenging obviousness and common opinion and reaching, no matter how, the expected goal.

The Tower has no actual use. This helps us, when observing the project, in considering that, whether "outer" aims, imposed or proposed by third parties, exist or not, the project still has an aim: the participants' satisfaction, the coming true of a wish, the realization of a dream. The project of the Tower of Babel, in spite of its tragic outcome, symbolizes a bond, and manifests the need of being and working together. A fundamental outcome of the project is, in any case, the well-being of having worked together.

Unlike the Leaning Tower, the project for the Tower of Babel doesn't reach a positive outcome, because mankind couldn't give up perfection. Perfection as absolute verticality, namely coherence with the idea of the perfect tower, is the idea the project expects to carry out. In the project of the Leaning Tower the important aspect is each single person's way of working, work considered as phenomenon: what shows itself and appears while one is working. But in the project for the Tower of Babel the work is *noumenon*, a work created out of the technician's mere intellect, a work turned into abstraction, work as supposed to be in the best possible world. So, the project fails.

In the Leaning Tower project, feasibility is not given beforehand: it's an emerging quality, is revealed afterwards, when, somehow unpredictably, the idea we started with clashes with outer environmental factors. The Leaning Tower project is carried out as adaptation, according to the environment: the result won't be the Tower we planned, but the Tower that may be. On the contrary, on the building site of the Tower of Babel, man let himself be carried away by arrogance.

Dante stresses the language factor: technical language, as intended by the technicians working at the Tower of Babel, is never redundant: the correspondence between container and contents is perfect. There's no noise, no sign without exact meaning, no distortion, no uncertainty. But this language doesn't exist, and the attempt to put it together and the illusion of possessing it are expressions of arrogance. Behind the perfect language lies the notion that there can just be one right way of working and just one correct and unmistakable way of expression.

But this is not how, in everyone's mind, and in the exchange between different minds, knowledge is constructed. Knowledge is constructed by approximation. Work is the product of experience, of trials and errors made one moment after the other. The Leaning Tower workers' experience compels them to change the project. Everyday experience can be taken into account, or overlooked. By being complexity oriented, we maintain it's worth taking it into account. Of all those involved in the project, each profession's adherence to its own technical handbook adds to the failure of the project. A single technical perfection doesn't exist, but rather several technical perfections, each coming from the different vision of the world carried by the different professional families involved in the project.

The project manager, the substitute for the absent God, remains far from divine perfection. As adapter, translator, creator of a context, ideal for none but livable for everyone as much as possible, the project manager bears witness to the awareness of the fact that better is the enemy of good, because where one sees perfection the other sees the flaw.

The project manager is therefore called to stand up for imperfection, redundancy, and noise. The project manager accepts and assumes responsibility for complexity, on the name and account of all the others, who prefer living in the illusion of technical perfection, all those who, not being able, and not wanting, to do otherwise, choose to look at the world from a single point of view.

TIME OF THE PROJECT

Should in technical language an approach to perfection be grasped, that would just be a photograph of language at a given moment. Language approaches perfection only if referred to that instant, at that particular moment. We can speak of a project either when we dream of something that doesn't exist, or in any subsequent moment and when we have an outcome. The complex notions of project and time are closely linked.

Time, in the logic of the machine/clockwork world, is thought of as a sequence of moments, of *attimi*. Literally, the Italian *attimo* means "atom," minimal unit, a "bit of time" not to be further divided. Empty minimal units of time, instants, one identical to the other, can be thought of as containers to be filled up with activities. This is one of the ways a project may be faced. One can look at it from the outside, as if overlooking a world one doesn't belong to, and describe what will be done in each instant between the instant considered the start of the project and the instant considered its end. So I make a list on a plane surface, on paper, of what every actor involved in the project will do moment by moment. The limit of this outline consists in its representing only one of the infinite number of possibilities. It describes only one of the ways of arranging activities in time. Things can go like this, but they can go in different ways as well. If, as a project manager, I devise this plan, and then try to make the life of the project fit my plan, I deny myself some opportunities. It could surely, from some other point of view, be suitable to arrange activities some other way.

Every instant alternatives arise. Every instant can be filled with activities in a different way. What seems from some point of view an indivisible unit of time (*attimo*), from another point of view appears as an instant: from the Indo-European root *sta-* (to stay), *in stare* (to stay close). Closeness in time derives from *instans* (what is happening right now), but in space too: what is close. The same attitude inducing us to plan, to the allocation of activities in minimum units of time, makes us understand instancing as classification. I give a definition of a hierarchical model of classes and afterwards I place the objects into the classes. This way we can instance on a huge board all the activities of the project. But we may grasp the same limits here we seize in planning: now, at this moment, I consider instancing in a given way adequate. At a different moment, things that now I see as far away, may appear closer.

Keeping faithful to a previously made instancing for the entire life of the project, I give up experience, the knowledge acquired by practice, one moment after the other. The project is actually a process, a never-ending becoming, a living system. I can "take its picture" every moment, and each picture will be different. An expression connected to instant and minute speaks exactly of process: moment. The process is a sequence of moments. A moment is an impulse, the contraction of *movimentum* (Latin for movement), from *movere* (to move), and from the root *meu* (to move over).

The project is a set of *attimi*, instants, moments. If I accept complexity, I had better consider all the moments and the instants equally important, because I'm not previously aware of what will be better. And anyway, the complexity of what happens, and the very fact that every stakeholder, including the project manager, influences the project by his observation, keeps me from knowing exactly not only what will happen, but also what happened and is happening.

CHRÓNOS VERSUS KAIRÓS

If I consider a project as a set of instants, I see the time dimension Greeks used to call *chrónos*: unobtrusive units that inevitably follow each other without breaks nor jumps. That is the time that flows in spite of ourselves, the time we're used to reading by means of clocks, calendars, and chronometers. The god Chrónos is the father of all things, and he devours his children. *Chrónos* is the time that devours man. The *chrónos*-time is an undifferentiated continuum flowing despite ourselves. The *chrónos*-time dictates the rules. In any moment we are subjected to its rule. In the *chrónos*-time there's no room for individual freedom. The calendar, the planning, the instancing determine what we have to do. In this time dimension—where any complexity is cleared away—project management is reduced to planning and control.

But the daily experience of those managing projects is never reduced to this. Because, in spite of the fact that the project manager is often forced to adore the god Chrónos, the time the project manager knows and lives through, is not *chrónos*, but *kairós*. Whereas *chrónos* is a time dimension that is imposed and suffered, *kairós* is a subjectively lived-through dimension; it's the dimension of experience. *Kairós* is the appropriate time, the convenient time, the "special" time, the good opportunity and the coincidence. Although from the *chrónos*-time point of view all the moments are the same, the *kairós*-time perceives the difference. For each thing you have to do, there's the right time, the propitious moment.

A unique *chrónos* submits everyone to the same rule, however, everyone has his own *kairós*: everyone has his own rhythm, everyone somehow enjoys individual freedom, and everyone undertakes his own responsibilities. (If I read "to have one's own rhythm" from the *chrónos* viewpoint, I perceive first of all a negative judgment toward those who go too slow; if I read "to have one's own rhythm" from the *kairós* viewpoint I mean "to move to the rhythm," "to feel the rhythm").

Kairós takes us back to two ancestral situations: the archer's gesture when he shoots the arrow toward the target, and the weaver's gesture, intent on making warp and weft correctly meet. In both situations our mind is tuned to seize the moment, the most propitious space–time window.

Looking at *kairós*-time, planning the use of time in detail appears useless; worrying to fill every moment of the life of the project with activities seems an idle thing to do. Plans and instancing remain in the background. Individual freedom seems irrepressible. And if everyone is stimulated to move at the propitious moment, instead of being squeezed on what belongs to him, it is only suitable.

In the *kairós*-time, you cannot say beforehand what shape the project will assume at a given moment. There will be quiet moments and moments of action, moments of idleness and moments of intense activity. The *kairós*-time has dealings with wisdom: some relative detachment, some inner tranquility is useful in order to read a weak signal, or to see clues, trails: to seize the *Zeitgeist*, the spirit of the time.

So the project manager, aware of the complex thin weave linking events, turns his attention to finding the less inadequate moment to do anything; less inadequate: anyway a suboptimal choice. In order to do everything, a still less inadequate moment than the one we were able to find could always have been found. But still an approximate one. It's idle to try to define, or fool ourselves into having found, the "right moment." More profitable, by hindsight, when all is said and done, is to go back to thinking of how that moment has been chosen, "by chance," through subtle sensations, by way of unpredictable coincidence. Even if we never find the "right moment," we can still improve our ability of intercepting the less inadequate time, our aptitude to "seizing the moment."

STATE OF THE ATMOSPHERE

And then an outer dimension has to be considered, one completely out of our control, but nevertheless definitely influencing the outlook of our activities, of our project: the weather, the climate. While we're busy planning and controlling, the sword of Damocles of outer imponderable variables, is hanging over each project.

The Italian word *tempo*, the French word *temps*, and the Spanish word *tiempo*, all stand for "fraction of a length of time," "sequence of moments," and for "weather." This wide spectrum of meanings doesn't let us perceive clearly the difference between the two concepts.

It wasn't the same in Latin: *tempus* only carried the chronological meaning, and the word used for "weather" was *tempestas*. In the Anglo-Saxon world the difference is quite clear. Time means "division," "sharing up." As the German word *Zeit*, its origin is in the Indo-European root *dai-*, hence in Greek *daiomai*: exactly "to divide," "to share out" (maybe *daimôn*, a "demon," too, in the sense of who shares out destinies). *Zeit*, time and tide speak of time only as "succession of moments," "a series of phases." In English we have, from this root, besides time, tide as well, nowadays meaning the "tide of the sea," but still used in the wider meaning of "period of time," or "season."

Surely seasons, as did lunar cycles, and astrological charts, always marked out to mankind the right time for the carrying out of this or that activity. We are bordering *kairós*, here the art (which should be brilliantly mastered by the project manager) of choosing the propitious time. But the fullest expression of complexity lies here, because we're still moving within a predictable frame. What is really difficult to seize is the sudden change in the atmospheric condition: precisely, a tempest getting closer. This explains the Latin expression *tempestas*: in the beginning it's the translation of the Greek *kairós*: "circumstance," "moment of the day." Later on *kairós* and *tempestas* assume, more and more precisely and technically, the meaning of "atmospheric condition," "atmospheric disturbance," or "bad weather." Expressed in English by weather and in German by *Wetter* (from the indo-European root *we-* "wind breath," from which also wind, Italian *vento*). That is the state of the atmosphere.

Whatever we're doing right now, we are inside an atmosphere, a "sphere of steam"; environmental conditions escape our will and our control. But through meteorology, "the science of the atmosphere," we can, somehow and approximately, forecast them.

ROBERT FITZROY: PROJECT AS A JOURNEY AND A SCIENTIFIC EXPEDITION

On the morning of December 27, 1831 the *Beagle*, a Royal Navy sloop carefully modified and re-equipped for the aim of the journey and the inclement atmospheric conditions awaiting it, sailed from the port of Plymouth, England.

Robert FitzRoy, captain and commander of the expedition, also the author of clever modifications to the ship, was a brilliant 26-year-old descendant of illustrious stock. His father, a general in the British Army, although family traditions indicated other courses, had accepted his son's childish dream: to be a seaman. This is why he has been in the Navy since the age of 12. He had already successfully commanded the ship in a previous expedition, but only after tough negotiations and political pressure had he been appointed chief of the present mission. To his advantage went the recommendation of his uncle the Duke of Grafton, member of Parliament, and an important representative of the Whig party, and the consideration in which Francis Beaufort, hydrographer of the British Admiralty, held him. Admiral Beaufort was the sponsor of the project, aimed at the cartographic survey of the South American coasts: the precise determination of the geographic co-ordinates of the ports, through the use of chronometers and astronomical observation, the use of the barometer, the examination of sea currents, and the experimentation of the scale, invented by Beaufort himself, for the empirical measurement of the strength of the wind, and also geological surveys and naturalistic research regarding the fauna and the flora.

On board, in addition to the experts' crew were the scientific expedition members: among others, a cartographer, a natural philosopher, an artist-illustrator, a missionary, and an enthusiastic 22-year-old, fond of geology, Charles Darwin. FitzRoy was well aware of the fact that solitude, extreme cold, tempests, and the desolation of the South Seas had brought Pringle Stoke, the previous commander of the ship, to suicide. This is why he had asked Beaufort for help in finding a gentleman companion, a young man to make conversation with, and with whom to share meals and the passion for scientific research.

From the discoveries made together, from the friendly and intense exchange of opinions, from the friendship tying the two young men, will the theory of evolution that will make Darwin world famous, emerge 25 years later. In the meantime, the friendship between the two had shown some cracks, because of differences in personalities, beliefs, and values. FitzRoy and Darwin had seen the same things, but had processed their experience in different ways. Darwin was a young seminarian with no perspective: FitzRoy's generosity and attention give him his life's opportunity. But later on, as happens, FitzRoy's star faded, while Darwin gained fame and approval.

The new paradigm of which Darwin is the standard-bearer teaches us that life is born out of chaos, in disorder: the idea of Nature as conscious divine project is rejected. And this is hard for FitzRoy to accept, although he had been making declarations about his availability to discuss any belief in the light of scientific evidence. We could compare FitzRoy's and Darwin's personal approaches before the project: while FitzRoy squandered his family's money and wasted a patrimony of high-ranking friendships in order to insist on projects he was convinced of, Darwin was minutely planning his career.

But I just look at the world from FitzRoy's side, which induces us to think that any project, if correctly understood, is a research project. The equipment, the scientific instruments, the supplies, everything is chosen and prepared for the long trip around the world, keeping in mind what is forecast will be done, and what is pictured may happen. But only later, cut off in the open sea, in unknown places not yet described in maps and trip reports, will what to do step by step be discovered. If the project is accepted in its complexity, it becomes research of all that can be done using those resources, moving in that context.

FitzRoy, as any scientific researcher and any captain and any manager and any project manager, could have kept his activity to a mere execution of his task, to the formal respect of a mandate, of a protocol. FitzRoy didn't disregard his commitments but, at the same time, inasmuch as he was traveling and had the opportunity of observing and having experiences, didn't waste the opportunity of discovering meanwhile what could be done (a project shows the part we are available to see) and thus goes to the core of what is the aim, the scope of his mission, beyond the content of formal documents, sating, in doing so, his own yearning for knowledge at the same time.

Beyond the visible borders of his own aims, out of FitzRoy's ability to consider the project a journey to the unknown, a road to the discovery of solutions, a research of answers to the questions that gave birth to the project, out of this open-mindedness, of this availability, the Beagle expedition brought back as a result an enormous unexpected progress in knowledge [5]. FitzRoy is also an emblem of a destiny I think every project manager has personally experienced. FitzRoy was the commander of the ship and the chief of the expedition. His was the economical responsibility, his the choice of people and equipment, his the risk, his the vision, the courage, the acumen, the everyday careful management, and the scientific knowledge. Thus was the new window overlooking nature opened. So the then penniless young man, Charles Darwin, got to publish in 1859, *The Origin of Species*. The bitter prize for all this is to be remembered as the captain in Darwin's journey.

ROBERT FITZROY: FORECASTING THE WEATHER

In the very years of Darwin's glory, we see FitzRoy engaged in his last project, a summing up of his human and scientific story. It also represents his drawing to conclusion his attitude toward project management. When he was 23, during his first experience as a commander on the *Beagle*, he met a terrible unforeseen storm on the Mar de la Plata. Aware of having been just a step away from shipwreck, FitzRoy had escaped it thanks to his heart and his ability of moving in uncertainty, at the cost of huge damage to his ship and its equipment and of irretrievable delays and emerging expenditures.

This experience is indelible and leaves its mark. FitzRoy will go on his whole life thinking that, had he been able on that day to sense the approaching tempest even one moment earlier, his reaction would have been more effective, the impact less dramatic, the damage less heavy, and the risk of catastrophe lower. Hence his interest in the complex system of marine currents, in the winds, the changes in atmospheric pressure, the groundswell, and the forming of clouds, therefore meteorology, the "study of celestial phenomena," and, more precisely, looking at the instruments and their practical application, the weather forecast.

Still supported by Admiral Beaufort, in 1854 he was appointed head of the Meteorological Statist with the Board of Trade. The Board of Trade, born as an inquiry commission, was responsible for the supervision of colonial affairs. We were at the height of the Victorian Age: fishing is fundamental for the nourishment of all Great Britain's inhabitants. The British Empire is a maritime power: it is based upon control of the routes and shipping trade. So the best naval engineering, the best navigation technology, and the best cartography, are needed. But later on one must navigate on the open seas, and live with the weather. And sometimes *tempestas*, the not-so-propitious atmospheric condition, may surprise us all of a sudden, a few miles offshore, as happened to FitzRoy that time in 1828.

Here is the strategic importance of the Meteorological Statist: something in between a service center and a study center, not for basic scientific research, but applied science and technology. FitzRoy and the three people working with him, equipped with a very small budget, worked in order to guarantee to people traveling by sea a reasonable knowledge of what the weather would be like.

BAROMETER AND READING OF UNCERTAINTY

In the beginning, FitzRoy worked on improving the barometer. He studied and adjusted a cheap standard model that had already been tested during the *Beagle's* journey—the storm glass—a sealed flask containing a mix of different ingredients: potassium nitrate, ammonium chloride, distilled water, camphor, and ethanol. If the liquid is clear, the weather is fair; if the liquid is cloudy, turbid, then the weather will be cloudy, perhaps with rain; if, on sunny winter days, the liquid contains some small stars, snow is to be expected; if crystals appear on the bottom, frost is on the way and so on.

It could seem a naïve search for analogies, but it's a powerful approach to applied research. There is no use looking for precision and certitude. Neither is it important for us to describe the passage from the water-based barometer to mercury-based and aneroid barometers. Here we just need to mention that FitzRoy was acquainted with all three of the technologies and that the important aspect for him was the function of the barometer, not its technology. It's an instrument allowing our minds to look for and improve the tuning with the situation, an instrument helping to perceive the condition of the atmosphere, and helping us to know somehow the unknown and the future. Galileo, Leibnitz, and Goethe did not all deal with barometers by accident (Goethe designed his own water barometer, called in German, *Goethe Barometer*).

The search for the way of moving in uncertainty seems therefore the richest manifestation of intelligence, a word literally meaning "reading between the lines," and induces man to create instruments. Instruments are not meant to substitute for man, but instruments may enhance man's ability to read clues. There's no breach between the ancient *haruspex*, extracting omens from animal entrails, or, precisely, scanning the sky, and what FitzRoy sets himself to do. The only difference stands in the accuracy and the effectiveness of the instruments he uses. So FitzRoy undertakes to equip the ships of the British fleet with barometers, provided with adequate instructions, and he supplies them with barometers at every port, even local ones from where the fishing boats sail.

WEATHER REPORT, WEATHER BOOK

This "reading the weather, the atmosphere," interpreting the clues according to experience, with the help of instruments, is the first step. Then FitzRoy moves to the processing of information. The indelible experience he lived there in the Mar de la Plata 30 years earlier combined with the impression caused by the Royal Charter Storm, the worst storm of the century, that lashed the British Islands toward the end of October 1859. Premonitions weren't scarce: "For a few days before the Royal Charter Gale came on, the thermometer was exceedingly low in most parts of the country: there were northerly winds in some places; also a good deal of snow; with low barometers" [6].

FitzRoy was aware that the storm could have been forecast, the sacrifice of hundreds of human lives avoided, and the economic damage minimized. So, exploiting the emotional wave caused by the event, FitzRoy arranged the service. A network of stations on the British coast and along the seas was already gathering information through barometers, thermometers, and eye observation. The information was used for statistics purposes. But now it was sent to the London central office through the telegraph. Here FitzRoy's and his three collaborators' active work registered everything on the maps.

> The first cautionary or storm-warning signals were made in February 1861. In August 1861, the first published forecasts of weather were tried; and after another half-year had elapsed for gaining experience by varied tentative arrangements, the present system was established. Twenty-two reports are now received each morning (except Sundays), and ten each afternoon, besides five from the Continent. Double forecasts (two days in advance) are published, with the full tables (on which they chiefly depend), and are sent to eight daily papers, to one weekly, to Lloyd's, to the Admiralty, and to the Horse Guards, besides the Board of Trade. [6]

I find the Italian expression "*annunci ai naviganti*" (announcement to seafarers) especially effective. "*Apri li orecchi al mio annunzio, e odi*" [7] ["Open thy ears and hear what I announce"]: the announcement is to make something known, to inform, but it represents especially a piece of news regarding the future: a prediction, a prophecy. The future, the atmospheric conditions that may endanger the project life, is unknown. But it may somehow be announced. Some clues are present and could effectively be used to read the situation we'll find ourselves in.

So in 1861 FitzRoy invented the weather report. The wireless telegraph was not available yet, therefore dispatches could not be sent to seafarers at that time. But ships won't sail if the atmospheric conditions appear too threatening. With the dispatches, the threshold of the risk implied in navigation is lowered. Nevertheless, due to lobbies' pressures (especially the shipowners of the fishing fleets, who considered information an inconvenience) the service was suspended. This contributed to FitzRoy's depression, and, nearing the age of 60, he committed suicide with his razor early one morning.

He left a book, published two years earlier. Reading FitzRoy's *The Weather Book* is for us a mostly interesting experience. It's a handbook. But it is a handbook supplying us with sure instruments. It's a handbook about how to move around in uncertainty. In the first page FitzRoy already observes wittily: "This book is intended to be popular—not necessarily superficial—but suited to the unpracticed and to the young, rather than to the experienced and skilful, who do not need such information" [6, pp. 1–2].

The recall to the pre-eminence of experience is only correct: no handbook will be able to tell me what to do now, in the present situation. But, as it speaks above all of a method, it actually is an instrument for experts, the method consisting of facing an invisible, yet impending, danger. FitzRoy deals with so complex a subject as the movements in the atmosphere, and we may infer, a subject as complex as our project is considered in the context of cultural, economic, politic, and strategic uncertainty: it is extremely difficult to combine mathematical exactness with the results of experience obtained by practical ocular observation and much reflection [6].

Only by "a means of feeling—indeed one may say mentally seeing" [6, pp. 1–2], can the dynamic laws regulating the phenomenon be perceived and then the propitious time for action be found. What is it all about then: not a method to be applied literally, but an attitude: the attention to what is happening right now, consideration for any useful piece of information, and the availability to doubt.

What is a weather forecast? "Prophecies of predictions they are not: the term forecast is strictly applicable to such an opinion" [6, p. 170]. It is something similar to an opinion, a conjecture, or a judgment based on clues of possible appearances. Not a refusal, but a special kind of scientific reasoning: we always have to think that the picture emerging from careful revelations, apparently coherent and stable, may become doubtful with a sudden wind or a stormy downpour we're given notice of by the barometer or by a mere glance at the sky.

CAPTAIN MCWHIRR READS *THE WEATHER BOOK*

I'm thinking of *Typhoon*, by Joseph Conrad [8], where the adventure of the *Nan Shan*, a regular cargo ship, in the China Seas during the typhoon season, at the end of the nineteenth century, can easily be read as a metaphor. Captain McWhirr, to common judgment, and to the eyes of his own crew as well, appears a "stupid man," of few words, normally gifted, yet reliable and suited to the task. Strictly traditional behavior is expected from him. Facing a typhoon is nothing new: sailing schools, handbooks, and acquired experience tell us what to do.

But we are not attending a training class right now: we are in the open sea. At sunset the sun "had a diminished diameter and an expiring brown, rayless glow," breathing is difficult for the sultriness, for the sticky heath, "a dense bank of cloud to the northwards" had "a sinister dark olive tint," and "low and motionless upon the sea," resembling "a solid obstacle in the path of the ship." The *Nan Shan* "after a pause of comparative steadiness, started upon a series of rolls, one worse than the other. ... The barometer is tumbling down like anything" [8, pp. 19–20].

Jukes, the chief mate, enters the chartroom and finds Captain McWhirr

[S]tanding up with one hand grasping the edge of the bookshelf ... reading a book. Now, here's this book—he continued with deliberation, slapping his thigh with the closed volume. I've been reading the chapter on the storms there. That was true. He had been reading the chapter on storms. When he had entered the chart-room, it was with no intention of taking the book down. Some influence of the air had, as it were, guided his hand to the shelf; and without taking the time to sit down he had waded with a conscious effort into the terminology of the subject. He lost himself among advancing semi-circles, left- and right-hand quadrants, the curve of the tracks, the probable bearing of the centre, the shifts of winds, and the readings of barometers. He tried to bring all these things into a definite relation to himself, and ended by becoming contemptuously angry with such a lot of words and much advice, all head-work and supposition, without a glimmer of certitude. [8]

Coincidences—the atmosphere directing the Captain's hand towards the shelf—are essential aspects of the time intended as *kairós, tempestas,* weather, propitious time, and special time. The state of the atmosphere is wind, the perturbation getting closer. But the atmosphere reigning on the ship, the thought forming in the captain's mind, is the perception of the exceptional event, of the impending breach in continuity.

So the captain comes across the meteorological handbook, read many times before. And the handbook opens by chance on the very page speaking of how to face a typhoon. But this time the words seem unrelated to the atmosphere. They look like advancing semicircles, quadrants: "With the diagram it may be readily seen that a current from SW, in the SE. semicircle, or half of a cyclone, impinged on by a NE. current, in the NW. half of a much larger and overlapping cyclone, must be rapidly, if not suddenly, turned (against the sun), or backed through SE., E., and N., to NE …" [6, pp. 56–69].

I am now reading FitzRoy's *The Weather Book*; I think that was the book Captain McWhirr had in hand. Let's try and figure out the situation: in a moment of illumination, in that special moment, in that very instant, in that sea, and feeling that particular typhoon getting closer, the captain finds the words of the handbook unsuited to the circumstances, to the atmosphere. Although he's familiar with them, the terms appear obscure, while he's reading here and there; he is lost among advancing semicircles, right and left quadrants, trajectory curves, the possible position of the vortex. He feels the need to get away from the book, and feels taking root inside the decision of taking action in a new way, unusual for him and for the crew and for the consolidated navigation experience. Instead of going around it, he chooses to cross the typhoon.

Rationalizing, we might say the motives of his choice were well founded because the usual procedure of going around a typhoon concerned sail navigation, and the *Nan Shan* was a steam cargo. But this is hindsight. At that moment Captain McWhirr was not taking that into consideration. At that moment the captain was out of the chronological time and living in the *kairós*, in the weather. He was at one with the storm. The important thing, what makes his decision, is the whole of the sensations he is feeling. At that instant the handbook indications and the barometer are but some of the sources from which the choice emerges. The choice that is forming in the mind and the oncoming typhoon are two aspects of the same "emergence." What matters is reading the situation and finding a suitable solution for the here and now.

Anyway, FitzRoy would have totally agreed with McWhirr. Keeping to the handbook at that moment would have meant giving up adequately facing the situation. The handbook and the barometer cannot and must not substitute for the feeling, the mentally seeing.

FORECAST

Fore indicates "in front of us," "facing us," and "in the prow," and *cast* means to throw or to fling. The fisherman hurls the fishing line, the archer knows the range of his arch, the distance his arrow can reach: forecast, to throw ahead, a gesture having something to do with the "broadcast," the wide sower's gesture, and the radio or television antenna covering with its signal a wide territory. But the idea of forecast doesn't refer to known territory. I know nothing, I cast my glance forward, I cast my heart beyond the obstacle, I imagine the future, and I cast a glance in a world that doesn't exist here and now.

Fore-cast and *pro-ject*: two closely connected expressions, and actions, and ways of looking at the world, to cast the glance and the thought ahead, and to imagine an action coherent to this glance and this thought. But important differences, too, are to be noted. The "traditional" instruments of project management are not imagined with the aim of forecasting anything. The vision, previously established, produced by the customer's mind, is assumed at the same moment of the mandate. Any idea of forecasting is even rejected: forecasting what will happen means no longer believing in the mandate. The project manager's consolidated instruments are not meant for forecasting; their aim is the comparison of what has been done with the mandate. And what's more, even where considering the *fore-casting* of the future, the forecast is considered an instrument intended for the finance domain: forecasting, yes, but only as to the required economic resources.

On the other hand, considering the forecast from FitzRoy's point of view, we issue forecasts without their being tied to a predesigned idea of what we are looking for in the future. The forecast throws us into uncertainty. But actually, whatever the mandate assigned, the world is what it is. We have no power of determining the occurrence. We don't know whether the ground will reveal itself similar to the ground underlying the Leaning Tower. And if, taught by the Leaning Tower experience, we invest time and resources in careful geological surveys, the outside factor, with a definite influence on the project, will emerge elsewhere, in some unexpected spot.

Our only alternative is therefore to forecast, to read the variations of the state of the atmosphere and the environment. The changes in the state of the context can be read as warning signals. Even accepting uncertainty, we still have the possibility, one moment after the other, of taking stock of the situation as a good captain would and measuring the distance to the port of destination, according to meteorological conditions.

After reading Captain FitzRoy's handbook and observing Captain McWhirr at work, the attention to outer changes required in every journey, in every project, becomes obvious. Attention to everything out of our control is still impending over the project. The best captain is not the one who rules the crew and organizes the work under regular conditions. The best captain is the captain who knows how to face storms. Saying, "It's not up to me," and considering ourselves not responsible for the consequences of the storm's breaking out, means not fulfilling our duty, the actual point of the project manager's role. It makes no sense either being fatalists or behaving passively, or trying to justify oneself, saying that there's no way of knowing when the storm is coming, or that each tempest is different from the others. FitzRoy is reminding us of the fact that what we don't know can somehow be sensed and be inferred from clues and weak signals.

The project is alive if, instead of just comfortably looking inwards, we look outwards. The project is alive if, one moment after the other, we face the weather, and anticipate, as much as possible, the incoming conditions.

In a careful examination, FitzRoy is moving the focus of our attention; he's inducing us to see a full picture where the void is, and compelling thorough reflection about our method, the project governance, and the use of the instruments. We owe him our contemporary awareness of how, in order for complexity to be faced, it's worth applying a wider look; dedicating all our attention to the management of the routine well-known world of our ship and crew is not enough. Considering the inner course of the project the main object of our care (as we used to do) is no longer advantageous.

Captain McWhirr was impeccable as to this aspect. He used to rule the crew with a sure hand; he was a careful manager of the load entrusted to him and of resources of coal. But he proves an excellent captain when he faces the storm in an unusual way. He reads the situation and finds a way of turning the catastrophic event into an opportunity; he disregards the rules, and instead of going around the storm, he cuts across it, as the steam navigation allows. He gets to the port earlier than scheduled, and saves resources as well.

What needs our close attention (we are not just happening to speak of soft skills) is the reading of the environment, of the potentially hostile atmosphere. The worse threats to the good course of the project come from the outside, from the sudden breaking out of the storm. Facing the storm is the trickiest responsibility of a project manager.

But on the other hand, we cannot allow the opposition of good weather versus bad weather to mislead us. The stake is not a simple linear choice. We are not talking about knowing when the bad weather will come, in order to lock ourselves up at home. We cannot even exactly define the "bad weather": in a place plagued by drought, "good weather" is represented by rain. We are talking, as always, of choosing the "propitious time." FitzRoy and McWhirr and every project manager must pursue their journey; we must continue working on the project. At stake is the timely understanding of the incoming weather, so as to combine as less inadequately as possible the activity to be carried out and the present weather.

We could even call every moment a moment of *tempestas*; *tempestas*, *kairós*, all mean propitious time. There's a less inadequate moment to do anything. In order to seize the "propitious time" for everything, we have useful "notices to seafarers," handbooks, and instruments for the governance of the project. And, today, instruments that are even more effective, being integrated ones: a personal computer connected to the Web acts as a barometer and receives the "notices to seafarers." The more powerful the instruments the more precise is the steering. But then, always, the interpretation and use of any instrument are still up to us. Neither the weather forecast nor the barometer tells us what to do. It's up to us.

A PAKISTANI PAPER MILL AND A POETIC, BUT TRAGIC, ACCIDENT

The Karnaphuli is a river in Eastern Bengal, the part of Bengal, now called Bangladesh, that became part of Pakistan, due to its prevailing Muslim culture, when in 1947 the new state was proclaimed and severed from India. Fifty kilometers from Chittagong, the first Pakistani port, overlooking the Bengali Gulf, along the Karnaphuli River, not far from Chandraghona, in an almost deserted area, the state body for industrial development, *Pakistani Industrial Development Corporation*, chose the location for what was to become one of the largest Pakistani industrial sites, one of the largest paper mills in the whole of Asia.

When the project took off, Pakistan had just reached independence: the new state was born in 1947. The borders between India and Pakistan were determined according to the prevailing religious culture. Eastern Bengal is Muslim, and becomes part of Pakistan. The project was based on the use of an easy-to-get raw material: the wide bamboo forests of Chittagong Hill Tracts, along the nearby tributaries of the Karnaphuli River. The plant began production in 1953. In the starting phase more technical and managing problems than foreseen were met, but when in 1959 the management passed to private hands (Dawood Group) the start-up phase could be considered concluded, and the plant fully operational.

Immediately afterward, nevertheless, a large-proportioned phenomenon threatened the very existence of the plant: the bamboo started blooming. It's a totally unforeseen event, and, according to the knowledge at that time and to the available forecasting instruments, a completely unpredictable one. Only when faced with evidence is the unexpected event carefully observed, and the plant scrupulously studied. The notion is acquired that the blooming happens every 50, or maybe 70, years. "In any event, the variety that supplied the Karnaphuli mill with some 85 percent of its raw material flowered and then, poetically, but quite uneconomically, died" [97, p. 9].

It was well known that the bamboo blooming would cause the death of the whole plant, and that regeneration would take place from the seeds instead of, as usual, from the rhizomes; what was not known was that the bamboo that had died because of the blooming wouldn't be usable for the wood paste because it would disintegrate during transportation, which was done by floating the canes downstream on the river waters.

Another unpleasant surprise was the discovery that, once the blooming ended, several years would be required before the new bamboo canes grew to a height that would allow their exploitation. So, in the seventh year of its activity, the plant had to face a very serious, and totally unforeseen, problem: an alternative raw material had to be found. As a temporary, and very expensive solution, the problem was solved by the importation of the required paste. But soon more original solutions were devised. An organization for the collection of bamboo in all Eastern Pakistan villages was set up: the river system crossing the country in every direction allowed cheap transportation; wood of any kind was cut in the plains. And, more important yet, a research program was started to identify new rapidly growing plants somehow able to replace the unreliable bamboo.

ALBERT HIRSCHMAN: ADVANTAGES OF UNDER-ESTIMATING

This story is told by Albert Hirschman. A student in Berlin, he was eighteen when he left Germany to escape racial laws. He studied at the Sorbonne, the London School of Economics, and in Trieste, where he found a master in his brother-in-law, Eugenio Colorni. In 1940 he sought refuge in the United States. After the war he worked as an economist in the Federal Reserve, then in Colombia, and was engaged in development projects. As a university professor, his approach was always distinctly interdisciplinary.

The case of the Pakistani paper mill led him to several observations. With hindsight, one could say that it has been a matter of "luck": "Its planners had badly overestimated the permanent availability of bamboo, but the mill escaped the possibly disastrous consequences of this error by an offsetting underestimate—or, more correctly, by the unsuspected availability—of alternative raw materials" [9].

Planning is abstract and cannot seize the specificity of the context. Therefore, the availability of bamboo was overestimated. But the error, the consequences of which could have been devastating, was compensated for by the underestimate (or, better still, by the totally unexpected availability) of alternative raw material.

This is an experience project managers share: because of the habit of intending the project as a plan designed beforehand, describing everything, one feels compelled to include everything in the plan, and then do only what is forecast. So resources are allocated beforehand in view of each foreseen activity; but that's not all: resources are allocated in view of every single risk. Therefore resources are taken up by activities that may not be necessary, swelling the cost of the project. Due to "luck," which is actually our consonance with the events when they happen, by chance a solution is found, but in the meantime resources have been taken up elsewhere to no avail.

Hirschman therefore wonders: because projects are usually, maybe unavoidably, described by mistaken plans, full of blanks, can it be reasonably believed that an association of errors systematically takes place, to be later providentially combined?

Expecting the overestimate of available resources to be always compensated by a symmetric underestimate of alternative available resources would be absurd. Nevertheless, the following general statement may be considered likely, even almost obvious: every project bears in its future several unpredictable threats and several unimaginable actions that could amend the threats. A general operative principle may therefore be formulated: the intervention of creative factors always represents a surprise; so they should never be relied on, nor counted on their manifesting, until this actually happens.

Consequently, in order to gather creative resources required for the success of the project, resources that may be available but that we cannot see, the nature of our undertaking should be misunderstood, and be imagined in a less exacting, simpler, and less demanding manner as to resources than what we deem necessary (that means: less demanding as to resources than what will turn out at the end as having been required).

Our fear, our anxiety, but our wariness too, make us underestimate our ability to give creative answers to emerging difficulties. Therefore, Hirschman suggests, we should also underestimate problems in the same way, so as to be led, by the converging action of two underestimations compensating each other, to complete tasks within our range, but that otherwise we wouldn't have had the courage to face.

PRINCIPLE OF THE "HIDING HAND"

Therefore hiding difficulties helps: if we don't cross our bridges before we come to them, the less we worry (beforehand), the more we carry on our work without anxiety, the more we will be willing to look at the world, and so find a way out, at the moment a difficulty appears. Hirschman calls this virtuous principle "the hiding hand."

We deal with the problems we think we can solve. So we end up coping with fewer problems than, if faced, would have been solved. But then, when uncomfortable or delicate situations appear, we are compelled to come face to face with problems. By then, in for a penny in for a pound, difficulties are faced, whether we like it or not. Because resources have already been invested and our prestige is at stake, the motivation for the identification and application of solutions is powerful. With hindsight, we can say that problems that (in cold blood, and when the dust has settled) would have been considered unsolvable (here's where the hiding hand helps) have been faced and solved.

So beyond a certain limit, beyond a general forecast, we had better avoid thinking of the future: the project that is there, and that may have a future, is the project before our eyes at this very moment. Difficulties requesting a certain level of commitment, if considered prematurely, can discourage the execution of the project, whereas they could be faced resolutely by assuming their responsibility subsequently, in the propitious moment. So, let's have recourse to "the hiding hand," especially in some phases of the project, even if partially conscious of doing so.

Hirschman dwells particularly upon the first phases. When a project is to be approved, people are spontaneously led to present it under a favorable light, because if the problematic aspects were underlined, doubts would be raised and the project would be unlikely to win approval. To obtain the approval of the projects, and to convince ourselves as well that they can be carried out, we use the "fake-imitation" technique, whereby we are led to sell new projects as imitations of successful projects already carried out, never mentioning the fact (here the "hiding hand" is at work too) that every project is different and that the possibility of reproducing already implemented solutions is not so certain as we like to believe. (The Leaning Tower proves it: the building techniques adopted for towers built on different ground didn't apply in that case.)

Another technique based on the "hiding hand" could be called, after Hirschman, the all-embracing technique. The different aspects of a project are all considered equally important, because, it's maintained, they are mutually linked. This is how an articulated, all-embracing plan is built beforehand: in this kind of project, the whole equals the sum of the parts. The accuracy of the plan allows the experts' ignorance as to courses to be taken to remain hidden, and offers an excellent alibi in case something goes wrong: the program will practically never be carried out to the letter; so the excuse "Things didn't go as they were supposed to because the program couldn't be respected," can be used.

Hirschman considers the use of these two techniques, "pseudo-imitation" and "pseudo-comprehensive-program technique," as "crutches to grope one's way" (p. 13) because we actually have to proceed blindly, for a good part. The project governance instruments aimed at helping proceed on an uncertain ground, even without an overall vision, even with a limp.

The "pseudo-imitation" therefore helps by making the project appear less fraught with difficulties; and the all-embracing program helps by beguiling us into possessing an all-embracing vision of the difficulties.

So the "hiding hand" actually assists us in facing complexity. But the price is self-deceit: it's an illusion, we act "as if" that program were realistic, even if we know deep down that such is not the case. On the contrary, complexity could be accepted with more awareness. The fact that the project, as a system, is not the sum of its parts, is not the mechanical assembly of single, clear, well-defined underprojects could be accepted. There's a difference between a global program founded on an overall perception of the "things to do" and program "smokescreen." The program based on overall perception, on the program that doesn't neglect complexity, is based on the awareness of the limit: our notions are not enough; clear thinking is impossible. Here is where the effort of limiting the number of converging actions to be undertaken should originate.

Meant and used this way, the "hiding hand," instead of being a constructive yet unwilling self-deceit, appears as a conscious minimization of requirements. If the ignorance of many aspects of the project were not hidden with plans and programs, but were instead accepted and shared, as Hirschman reminds us: "the indications would converge towards the few measures we can adopt" (p. 17). After all, the "hiding hand" avoids the description of complexity on paper. Regardless, it is useless, maybe impossible, to describe complexity. Complexity can only be lived through. Once the providential hand hides difficulties, we could move step by step, trusting instruments that allow us to look at our goal, to see how the situation stands, make the most of the resources, and mitigate the negative impact on the environment.

WHAT IS AT STAKE?

Stake, originally, "a post or a stock planted in the ground" comes from the Indo-European root *steg-*, the Proto-German *stakon*, whence derive the Spanish *estaca*, the ancient French *estaque*, the Italian *stacca*, the ancient English *staca*, then in modern German and in English, *stake*.

From the pole to the identification of the place where the pole is planted in the ground, and where gamblers assemble, there, on that very spot, the "stake," the total of the money collected among the bettors, is piled up. While waiting to know who will be the winners, waiting to know how it is going to be divided, the sum must be taken care of, and for this someone trusted by all players is required. This person will soon have a name: stakeholder.

The stakeholder therefore answers an interest held by all players: he is the "guarantor" that the rules of the game are kept, and he is the guarantor of the equal sharing out of the stakes. Hence the ancient Anglo-Saxon legal institution: the stakeholder is the "trust manager," the administrator of property belonging to a third party prevented from personal management, for instance, by minority or public office.

The stakeholder, therefore, acts on account of other subjects and looks after someone's interest. The Latin verb *interesse* stands for *inter* (between) and *esse* (to be), to be between this and that, exactly in the way the manager of the stakes stands. The Latin verb implicitly refers to the world of organization and work: *interesse negotia alicuius*, to be "between someone's things," to be involved in someone's business. Seen from the stakeholder's point of view: another person's business works if I am involved; with my intermediation third parties can reach their aims.

Thus the stakeholder, the person allowing other subjects to reach their aim, is originally the project manager. All stakeholders, or better yet, all joint partners involved in some undertaking, in an activity, in a project, entrust the project manager–stakeholder with the resources at stake, and also put their hopes, their dreams, and their expectations for a future outcome in his hands.

Later the expression evolved toward today's meaning: everyone is a stakeholder, everyone possesses his own stake, his wealth, and his own values to risk. But the memory of the original meaning is still present in English (although lost for other languages' speakers who just borrow the English word): my stake has a meaning only as part of a shared stake, of an assembled stake; stake means involvement.

In the language of finance, nowadays, the shareholders' interest, belonging to the holders of financial resources, is considered the prevailing one; other stakeholders' interest, the interest of others involved in an enterprise, workers and other subjects, just comes after. But the etymology is a reminder of the fact that today's shareholder was called stakeholder, and that a shareholder is but one of the stakeholders. We can maintain that if we want any undertaking to work, all stakeholders must be taken equally into account. The same can be said about the interest. Everyone bears "his personal interest" and acts in "his own interest." But according to the history of this expression no one can pursue his own interest alone, without taking others into account because—*inter esse*, "to be between"—we all have the other under our feet; we unavoidably trample on the same ground.

If this is true in general, it is especially true for the project. Only if all players deposit their money near that pole, in that shared place, only if everyone puts his own resources at stake, can the goal of the project, of something actually going beyond what I already know, I already have, be pursued. These collective resources can be directed toward the shared goal only if someone, in everyone's name, undertakes to hold, as far as possible, everything together: we once more go back to the project manager.

LUMEN VERSUS LUX: STAKEHOLDERS, CONES OF LIGHT, AND PROJECT ETHICS

When the project manager looks at a project, and divides everyone's activities in units of time, he is dividing the activities as he thinks is better at a given time. In another moment the job would have been done differently. And not only that: his way of dividing activities is only his own way. What is being said is being said by an observer. As the project manager looks at the world, so does any person involved in any way in the project. The project manager is but one of the actors, one of the stakeholders.

The conception we are dealing with is an essentially plural one: there are a lot of stakeholders, each bearing his or her own point of view of the world. So, considering stakeholders are all the subjects touched on by the project, all the subjects that touch on the project should become common ground inasmuch as they work on it or because they are interested in its outcome. The people with a negative or positive influence on the project are counted among the stakeholders, of course, but are not the only ones. All stakeholders are equally important, not according to an abstract equality principle, but because everyone affects the project, whatever his contribution in resources and facilities or difficulties. Should a stakeholder, even an apparently marginal one, be missing, the project would be different.

We are all assembled, as gamblers around that pole, in a single virtual place. But everyone, a different person, the bearer of a different story, a different culture, a different aim, casts a different look on that place. Everyone observes that world with different eyes and sees different things. Each person projects on the world his own cone of light. As a consequence, the project that is there, the project that can be seen and considered as shared interest, is in the place on which all glances converge, the place floodlit by the different cones of light borne by each stakeholder. The project that can be carried out is the area toward which the cones of light converge, that is, the site of the future on which all stakeholders agree.

Scholasticism helps explain the matter more clearly: light can be intended in two ways. There are two ways of accumulating experience, of building knowledge, or, generally speaking: planning. There's *lux*: it's the light shed by what has been done and by the plans designed. The *lux* is the picture, the image, and the representation. There's a saying "in black and white": a contract has been signed, tasks have been assigned, engagements have been undertaken, and everyone is confident. Then the advancement of the project is described according to what the project should have been, so the *lux* of the Leaning Tower project spreads over plans and illuminates a perfectly vertical tower that couldn't and would never exist. This is probably why these written traces have been cancelled, or forgotten: too far from the project that could be seen.

The concept of *lumen* is different: *lumen oculorum* (light of my eyes), *lumen intellectuale* (light of the intellect), and *lumen fidei* (light of the faith); lumen means "inner light," the "light emanating from the eyes," from a person's mind and heart, from every person, and as it rests on the world, it lights it up.

Accepting the project's complexity means accepting each stakeholder's, each involved person's, different light. The project, a living thing, is constructed by stakeholders' converging glances. Each stakeholder's *lumen*, one moment after the other, lights up the darkness, so under the light of that *lumen*, of each stakeholder's different *lumen*, one moment after the other, the project takes shape, and any representation becomes useless. Every stakeholder gives his contribution: he seizes the moment differently, and so sees close connections between different things.

The *lumen* boring a hole in darkness builds the project when *lux* is absent as well. So, in the place where the cones of light intersect, where the glances converge, the project is a fantasy taking shape. Fantasy's root is the Latin *phantasía*, from the word *phanein*, "to show," "to make visible, obvious" (hence *epifania*: the "feast of the appearance"). The *phàntasma* (Ancient Greek for phantom) is the "appearing image." The root is the Indo-European *bha-*: "light," "illumination."

Therefore we can consider the project as superimposition and coincidence, or reasonable convergence of *lumen*, glances, cones of light. We can say, with Emmanuel Lévinas, that the project is a "vision without image," a vision existing whether reflected in the mirror of some document, or not. Lévinas, to be precise, is speaking not of projects but of ethics, and he says "ethics is a point of view" [10]. It's our point of view on the world, deep-rooted in our values, our culture, our personal history.

First of all stands the self. Everyone, every stakeholder, every "bearer of interest," is different from any other, and, maybe, is at the same time, "identical," and "exactly the same," but only to himself. He is an unrepeatable being, with his own, relative, point of view (his *lumen* doesn't see everything) but matchless. At the origin is the Indo-European root *s(w)e-*, "oneself." Hence we have in Latin *se* and *suus*, in Italian *sè* and *suo* (himself and his), *soi* and *sien* in French, *sí* and *suyo* in Spanish, *Sich* and *Sein* in German. In Greek *hos* means "he" or "this"; *hekastos*, "everyone." From here, we go to collective identity with *ethnos*, where *êthos* speaks of "individuality," "what belongs to one person," "custom," or "way of life." Everyone has his own ethics; ethics is a point of view, and the project is a point of view.

No actor, no stakeholder, can bring a project to a successful conclusion by himself. No actor, no stakeholder, can be excluded from the scene. And when he is present, better remember it. Even if I consider him hostile, he will still observe the project from his point of view. The project exists because different points of view exist. The project is the meeting point of these points of view.

ATTITUDES BEFORE THE PROJECT

The construction of the project depends on the attitudes of the people involved. We can stand before the project in different ways. Actually, as it happens, in different moments one lives through the project differently: anxiously, fearfully, joyously, or considered a bother. And this happens to all of us. Therefore, a peculiar aspect of complexity in projects is that a project is what different people see, and their attitudes change in time.

To try to make what I mean clearer, to speak about multiplicity and about the plural nature of a project as seen with the eyes of those involved, I use an interpretation I infer from my ethics. My personal story, my culture, are the sources determining the way my point of view, my lumen, one of the many possible lumen, manifest themselves.

This interpretation can be applied to the attitudes of all people involved in the project. I know the role of project manager is different from the role of every other person involved: but he is a *primus inter pares*, the difference doesn't consist in his skills, in his professional role, but in his ethics: a project manager's role is to keep in mind the various kinds of *lumen*. His, more than anyone else's, is the responsibility of identifying the shared area illuminated by the various cones of light.

WORKING ON THE PROJECT MEANS BEARING ONE'S CROSS

The Latin word *labor*, where the Italian *lavoro* (work, labor) comes from, suggests the idea of a hard and painful activity. This idea probably comes from the verb *labare*, "to stagger under some weight." In French, Spanish, and Portuguese too the words (*travail, trabajo, trabalho*, respectively), tell us a story of suffering. They come from an instrument of torture, made, according to tradition, by three poles (in Latin *tripalium*), to which the offender was tied. The idea of work is inextricably linked to crucifixion.

So is any work. But the project implies a greater assumption of responsibility compared with what happens in subordinate employment, with a task, or with a kind of organization. The very nature of a project denies room for escape or evasive behavior; the possibility of hiding or relegating one's burden to others is much smaller than in regular subordinate work: from all roles just one behavior is expected: contribution to the result. And the expected result, contrary to what happens in subordinated work, is obvious for everyone.

Everyone working on the project is asked to bear weight, to assume responsibility. Everyone lives through his ordeal and suffers injustice. On everyone's shoulders falls the weight of a cross, of the same cross—fate's supreme irony—to which he will be nailed if things don't work out in the right way. At the same time, the cross everyone bears is the means of salvation for the others. The common goal compels everyone to consider that the other's cross belongs to him as well. So "bearing one's cross" becomes a service. This is true for anyone working on the project, but it is especially so for the project manager.

The project manager is not endowed with the boss's stripes: he is a mysterious character, whose duty gets specified day by day, to whom an important role is entrusted, a decisive one for the story of the project: his role is to be of service to others. If it's true, as it is, that service comes from servant, does "offering one's service" mean putting oneself in a subordinate position, intrinsically tied to the servant/master dialectics? Not so: after all, the Latin word *servus* did not originally mean "slave," but "shepherd."

Service comes from the Indo-European root *swer-*, expressing a wide range of concepts: "to see," "to look, to keep, to give guarantee, defense, or protection." We derive from *swer-*, in Sanskrit *varutà*, "the protector," in Greek *horán*, "to see," and in Latin *observare*, "to observe": *ob*, "towards," and *servare*, "to keep," with the double meanings of "to be careful," "to fulfill," and "not to avert the eyes from." The service is connected to vision, the *lumen*, wisdom, and knowledge.

As we read in Chapter 42 of the Book of Isaiah [11], about Jahweh's servant, a project manager:

> He shall not cry, nor lift up,
>> nor cause his voice to be heard in the street.
> A bruised reed shall he not break,
>> and the smoking flax shall he not quench.
> He shall bring forth judgment unto truth.
> He shall not fail nor be discouraged
>> till he completes the project.

WORKING ON THE PROJECT IS LIKE BEING GOD

The project manager, as we saw, is the substitute for the absent god. He cannot guarantee the completion of the project of a tower reaching up to heaven. But he can guarantee a project not tainted with arrogance to terminate with some kind of outcome: the tower, as in Pisa, a leaning one and far from the plan designed beforehand, yet endowed with intrinsic value, can be built.

Arrogance means an exaggerated claim. Ambition is burning desire. Whoever understands the project rightly is ambitious, feels something is missing, and wishes to fill the void. He wants to go beyond the limits of what can be seen, he challenges the unknown, and he speaks about what is not yet there. He is a prophet, he announces, and he predicts, a very different attitude from planning and programming. The prophet's word is creative and performing: a word acquiring its meaning and keeping its ethical dignity only if transformed into performance, it speaks about an action and motivates to action and finds response in action only if it's a word aimed at bringing change, words thought with just a single aim: creating a world. A word speaks of a world that isn't there yet, in order to make it exist.

> For, behold, I create new heavens and a new earth:
>> and the former shall not be remembered,
>> nor come into mind.
> But be ye glad and rejoice for ever
>> in that which I create. [12]

Whoever, in any time and any place, during the presentation phase or the "progress reporting" phase, speaks about the project, should bear this clearly in mind: the words being pronounced do not describe the project from outside, but are part of the project itself and contribute to its construction.

Sometimes, on the other hand, words are used to hide from ourselves and others our inability to see where we are going, the fear of complexity, of the darkness in which we are moving. But these are words blowing in the wind, empty and useless rhetoric, that don't distance us from the perception of problems, and don't contribute in the least to their solution.

Along the difficult life of the project, we often feel compelled to speak in order to hide, deceive, or stall others. We should on the contrary think that we can use words to persuade, convince, and seduce: everyone's attention and skills would then be focused on the goal. Prophetic words remind us that what we say, even the words dedicated to a world we can hardly see, can shed light. These words are closely connected to our eyes, to our lumen.

Creating, a situation totally different from the execution of something foreseen, means considering chaos constructive: in chaos, in obscurity and darkness, is the world born. Modern science deep down speaks of how the Tower of Babel could not be built: claiming to dominate and control life is arrogance. Life is not to be subjected to some project of ours, but has got its own implicit project to which we are all called to contribute. Here is what the role of project manager is about: in a world with no God, or one abandoned by a God preferring to leave us alone in the face of our responsibilities, we are able, step by step, to cast light in darkness.

In the project's chaos, in the project's daily life, where the highest disorder reigns despite all plans and programs, there is the very place where the roots of an emerging world can be seized. The words of those governing the project, and their actions, cast light over the scene and look fearlessly at chaos. They untangle the knot and direct the work toward the goal.

This is what a good part of the project manager's work means, in the same way as the prophet denounces what is wrong and announces the aim moving us, in a way suiting each moment. What is true for the project manager, is true for any person involved in the project. Everyone is required to contribute to the creation. The project is a collective delivery. "Delivery," too, means going from darkness to light. Let's keep in mind all expressions speaking of coming to the light.

WORKING ON THE PROJECT MEANS BUILDING ONE STONE OVER THE OTHER

In a project, careful and slow construction cannot be skipped. Construction means working carefully, and being aware of details. Construction means working with care. The idea of "care" takes us back to "concern": one "takes care" of the project, the project "concerns" us, and the project "relates" to us. The project concerns, and relates to, all people involved. But it concerns the project manager, first of all, and relates to him; he testifies, on everyone's account, to the possibility of reaching the result. To concern implies "mixing together." "Mixing" means forming a single mass, flowing in a whole, forming an homogeneous whole. Only if from the resources a good mix emerges, may something good come out of it. This is on what the project manager, and whoever works on the project, with any task, should concentrate.

Building means, again, concentrating on structure: on correlations, on interdependence, and on connections. Therefore the explicit talk between architects and stonecutters and masons Dante mentions is fundamental. Building one stone over the other means keeping in mind that a project is built by laying the basis, starting from the foundations.

> Behold, I lay in Zion for a foundation a stone,
>   a tried stone,
> A precious corner stone, a sure foundation:
> He that believeth shall not make haste. [13]

But the question is not looking for abstract perfection. The attachment to a model, to a plan, to a "should be" of the project, is the understandable consequence of our fear of not being in a position of coexisting with complexity, and yet this attitude takes us away from the result. The Leaning

Tower has been built one stone over the other, step by step, getting away little by little from the idea of a tower rising vertically toward the sky, and adjusting the project little by little to the state of the world.

We must build using the stones we find. This is how Romanesque cathedrals were built by reusing the columns of Roman temples. The best stone ever is the one available while I am working. The construction I can complete is based on these "second-best" stones.

> The stone which the builders refused
>   is become the head stone of the corner. [14]

"Building one stone over the other" means "putting together step by step." It's a progressive refinement. In order to build the Leaning Tower, 300 years, and all the technological discoveries made in the interim, have been required. But any Factory of the Milan Cathedral [proverbial for "never-ending job," translator's note] lasts hundreds of years, and is still open after the church has been consecrated. Of any project, observed in its complexity, the starting point, the moment in which the foundation stone has been laid, the Big Bang, can be determined, but not the closing point.

The closure is a convention: an agreement between all the people involved, and a declaration of acceptable satisfaction. The choice of the moment for release and the passage to the maintenance phase is an agreement. But the work, from the point of view of constant improvement, is never finished; a stone must always be added or replaced. Nowadays, for instance, in the light of development technologies allowing the development site to be considered permanently open, and the involvement in the work of the people once relegated to the passive role of final users, a software development site is permanently open: this is the subject of the permanent beta. A project is therefore accepted as never completed: the result is always a temporary one, defective but improvable all the time. The project has an end but no ending.

WORKING ON A PROJECT MEANS CASTING BREAD UPON THE WATERS

"Cast thy bread upon the waters: for thou shalt find it after many days," says Qohelet [15]. A project is built by displaying day by day confidence and reliance, self-respect and consideration for others, and peace of mind.

A project is built by shedding illusion and attachment to any source of reassurance. Although I keep tied to my plans, the project, a living system, grows as it can and as it wants.

> To every thing there is a season and a time to every purpose under the heaven. [16]
> A time to weep, and a time to laugh;
> A time to mourn, and a time to dance;
> A time to cast away to stones, and a time to gather stones together.
> A time to get, and a time to lose;
> A time to keep, and a time to cast away;
> A time to rend, and a time to sew;
> A time to keep silence, and a time to speak. [17]

There's no reason, no model that can tell us which is the "propitious moment" to act, to do one thing or another. Nor does reason help reading weak signals, seizing the moment when the storm is approaching. Therefore, to let the project live, and do so by minimization of requirements and limitation of interference, wisdom is needed. For a project manager, and whoever works on the project, wisdom is more important than reason. Wisdom is moderation and balance and means a knowledge of things acquired through experience. Therefore, we have "cast bread upon the waters," accepting the project that "makes itself," and observing the project grow as one would observe the flock, or look at the wheat growing.

CONVERGENCE, PRESENCE, EXPERIENCE

The project lives in the place where glances and attitudes converge. Several subjects are looking at that spot, that place which little by little ceases being dark, that clearing in the forest. The convergence of glances, of everyone's *lumen*, casts light. The convergence of attitudes: bearing one's cross, being like God, building one stone over the other, and casting bread upon the waters, all combine and make up as a whole an effective behavior: an existing behavior, a behavior manifesting itself moment by moment, but one eluding any control.

What is being said is said by an observer; we shouldn't forget our point of view is always relative and incomplete. The same can obviously be said of the most important sponsor's and stakeholder's points of view, and of the project manager's point of view, as well. We cannot look at the project from the outside. We should remember that every time we plan and monitor, every time we check, and every time we just observe what is happening in the project's life, we are influencing it. The observer is part of the object of his investigation. We cannot objectively investigate reality outside the relationship it holds with the subject living the experience. We may say the project is a muddle we are part of, a muddle sorting itself out under our very eyes. Our contribution consists in making this sorting, this inclination toward a solution more effective.

I think this represents sorting out the muddle: putting my attention on my way of constructing knowledge at this very moment. The useful knowledge is the one emerging here and now. The already given, judged and judging knowledge relates to the past. Experience is obviously useful but it precisely consists of applying what one knows to the present situation.

The way of acting that can be called scientific, the same way of acting of the IT professional working on pure, granular, nonredundant, data, this scientific way of acting eludes the key question, the question a project manager should ask himself constantly: "What is happening?" What is happening is the sum of the experiences all the people taking part in the project lived through.

TOOLS FOR PROJECT GOVERNANCE

We had been working for more than a year at examining the project under the light of complexity, when I was asked the question I consider the major one. What difference is there in the instruments on which we are working? To try to answer this question, going back to a topical moment in the history of Western philosophy has been helpful, and still is.

Kant in his *Critique of Pure Reason* [18] wonders how our knowledge is constructed. His key idea is synthesized in the preface to the 1787 edition. Most of the scientific method and the mathematics of the following two centuries are descended from this idea. From there management is descended too, as codified since the 1930s, and all computer science based on analysis and data modeling as well.

Kant writes: "Reason sees only what it produces by itself according to its own design" (p. 106). Reason must "get in first and compel nature to answer its questions without letting itself be guided by nature, because otherwise our observations, born out of chance and without a pre-designed plan, wouldn't result in a necessary law, that nevertheless reason seeks and needs."

Kant is saying to us: knowledge is usable only if kept together by an idea (under an idea) co-ordinating its single parts, and granting each part a specific place. Reference to a preliminary criterion is required, independent of experience, empirical data, and beyond that set of notions: a transcendental criterion, above, and aimed at organizing, that medley of information. "I call transcendental any knowledge not dealing with objects, but with our way of knowing objects, as this knowledge should a priori be possible."

In short, as acknowledging phenomena, "what shows and appears," is too difficult; according to Kant we had better rely on reason. What reason sees is not things, but abstract representations of things, *noumeni*, "what is being thought." As Kant says, inasmuch as we are not able to see, to grasp the project in its enormous complexity, and we are not granted the possibility of dominating and including everything in our glance, we choose not to look at the phenomenon, at what is nevertheless before our eyes, but to give all our attention and our care to a rational image of the project.

Following Kant, we can plan and control projects, but at a hardly negligible price: that of substituting the project, a living thing, the project as desire and meeting place of the glance, with its cold representation. Every project manager is aware of this; every project manager personally experiences this annoying feeling of detachment from reality.

Kant, after all, is innocent. He wrote more than 200 years ago, in a different historical, cultural, and philosophical context from the one we are moving in today. He synthesized epistemological reasoning that had been uncertain and muddled before he came. After all, he didn't stop wondering, always keeping in mind those bearing different opinions, such as Hume, an empiricist, who was skeptical of any method, and always curious toward Swedenborg's parapsychological talent: a scientist, and an inventor, he later on turned to theology and the project of a new religion.

On the other hand, all the scholars who used Kant's philosophy and transformed it into a school, should not be considered as innocent. A dangerous school, because through a method substituting a project with an image, it opens the way and offers legitimization to the claim of subordinating the project to a superior outer control. Now, there's no reason to deny its value to Kant's method. But still, in our opinion, this is the way many projects fail. If an effort is made to bring back one minute after the other the Leaning Tower project to its representation—a straight tower outstretched toward the sky—the project's failure is decreed.

So, during the first years of the twentieth century, going back to Kant's questions and going beyond Kant, and under the influence of the new approach to knowledge new disciplines, such as psychology, psychoanalysis, ethnography, and epistemology, the method we adopt to access knowledge, is drawn back to the center of attention. While Kant was proposing a single epistemology, now the thinking is of the contemporary presence of several epistemologies. Everyone has his own epistemology, every person involved in the investigation, every native observing the world beyond his own culture, and every stakeholder involved in a project. Every epistemology produces a story, different from all others.

What matters is the phenomenon, what is happening, what I am able to see, here and now. Once more, we are speaking of gazes. The scholastic opposition *lumen* versus *lux*, from the systemic framework enforced by Middle-Ages philosophy, is up to date again. Husserl speaks about pure look. The purer the look, the less it is conditioned by preconceived ideas and prejudices, and the more profound and richer will knowledge be [19, Lecture II, p. 106].

This is why, instead of reading each project through a single articulated (and subtle and refined as it may be) epistemology, we should consider that every project brings along its own epistemology. As each project is different, as each project is the result of particular gazes focusing on one goal, the epistemology suiting that project emerges from the project itself. It is up to us, mainly to the project manager, to see it and use it.

Here is what I have been trying to say in my answer to the question. Why if we look at a project and accept its complexity, do we need different instruments from those we usually apply? And what exactly is the difference? I have no answer. But maybe I can give an indication of the journey the expert project managers I have been working with followed. The second part of this book is the story of this journey. Here and now, I only draw the conclusions of my argument.

> If we accept the project's complexity, and we choose to live through this complexity in full awareness, we should at least think of representations—images, pictures summarizing the project—evolving in time, adapting to the environment.

If we accept the project's complexity, we should to try to devise not instruments of control, a comparison with the a priori, the beforehand, but instruments of governance, instruments to move in uncertainty.

If we accept complexity, we should think that every instrument will be able to take a project into consideration only as to one aspect, only from a given point of view. In the optics of complexity, no instrument can be generalized. I must expect the tool to reveal itself inadequate at all times in the face of the emergence of an unexpected context.

Faced with complexity, every instrument marked by a way of using that was defined beforehand, is not an effective instrument. Here, a metaphysic vision of the instrument has been assumed as the starting point: its aim is given beforehand; we only have to learn how to use it.

Heidegger helps us see how ineffective this attitude is and how vain from the point of view of the person at work. Only I, here and now and being in the situation, can be aware of what I need. Therefore, maybe, more than with new, or additional, instruments, we should equip ourselves with a toolbox allowing us to prepare an adequate instrument each time. We should then speak of meta-instruments, or deuteron-instruments. The object in itself has no value, the value belongs to the object tested in everyday life, in a single situation. Or better still, the generic object becomes in our eyes a finalized instrument the moment we use it.

In Heidegger's language this is *Zuhandenheit*. To give an idea of how wide this concept is, I quote translations suggested from German to several other languages: *readiness-to-hand, handiness*; *essere alla mano, utilizzabilità* (Italian); *ser a la mano, ser ante los ojos* (Spanish); *utilisabilité* (French) [20].

The specific "handiness" of the hammer is revealed by the hammering itself. I think, I hope, a project manager can avail himself of similar instruments, instruments born out of specific needs found in single projects, instruments such as you will find suggested in the following pages, instruments such as compasses and barometers, "notices to seafarers": instruments that are helpful in concentrating one's attention on the moment and the uniqueness of the situation through which we are living, and instruments that present themselves exactly as aids to move around in uncertainty, to live through the present moment, and seize the propitious time.

REFERENCES

1. Gadda, C.E. (1957). *Quer pasticciaccio brutto de via Merulana*. Milan: Garzanti. [English translation: *That Awful Mess on the Via Merulana*, New York: New York Review Books Classics, p. 5.]
2. Calvino, I. (1988). Molteplicità. In *Lezioni Americane*. Milan: Garzanti. [English translation: *Multiplicity*, in *American Lessons*, Cambridge, MA: Harvard University Press, 1988. p. 106.]
3. Carli, E. (Reviewer) (1989). Il campanile. In *Il Duomo di Pisa*. Florence: Nardini.
4. Dante, A. (1303–1305/2005). *De vulgari eloquentia*. In Mirco Tavoni, (ed.) Dante Alighieri, *Opere*, volume I. Milan: Mondadori, 2011. [English translation: *Dante: De vulgari eloquentia*, Steven Botterill (ed.), Cambridge Medieval Classics, 2005.
5. FitzRoy, R. (1839). *Narrative of the Surveying Voyages of His Majesty's Ships* Adventure and Beagle *between the Years 1826 and 1836, Describing Their Examination of the Southern Shores of South America, and the* Beagle's *Circumnavigation of the Globe*. London: Henry Colburn.
6. FitzRoy, R. (1863). *The Weather Book: A Manual of Practical Meteorology*. London: Longman, Green, Longman, Roberts & Green, pp. 1–2, 171–172.
7. Dante, A. (1304–1307). *La Divina Commedia,* a cura di Giorgio Petrocchi, Milan: Mondadori, 1967.
8. Conrad, J. (1903/2003). *Typhoon and Other Stories*. London: William Heinemann, 1903. Now published as *Typhoon and Other Tales,* New York: Oxford University Press. 2003.
9. Hirschman, A.O. (1967). *Development Projects Observed*. Washington DC: The Brooking Institution, p. 9.
10. Lévinas, E. (1961). Totalité et Infini. *Essai sur l'extériorité*. La Haye: Martinus Nijhoff, Préface, p. XI.
11. Bible, Book of Isaiah, 42: 2–4.
12. Bible, Book of Isaiah, 65: 17–18.
13. Bible, Book of Isaiah, 28: 16.
14. Bible, Psalms, 118: 22.
15. Bible, Ecclesiastes, 11: 1.
16. Bible, Ecclesiastes, 3: 1.
17. Bible, Ecclesiastes, 4: 7.
18. Kant, I. (1787/1998). *Kritik der reinen Vernunft. Zweyte hin und wieder verbesserte Auflage*. Riga: Johann Friedrich Hartknoch, 1787. English translation: *Critique of Pure Reason*, Paul Guyer and Allen W. Wood (eds.), Cambridge University Press, 1998, Preface to the second edition, p. 106.
19. Husserl, E. (1907/2010). *Die Idee der Phaenomenologie. Fünf Vorlesungen*. English translation: *The idea of phenomenology*. William P. Alston and George Nakhnikian, (eds.). Boston: Kluwer Academic Publisher, 2010. Lecture II, p. 22, 106.
20. Heidegger, M. (1927/2001). *Sein und Zeit*, Niemeyer, Tubingen.18th edition 2001. *Drittes Kapitel: 'Die Weltlichkeit der Welt' (the Worldhood of the World), 'Die Analyse der Umweltlichkeit und Weltlichkeit Überhaupt'(Analysis of Environmentality and Worldhood in general)*, § 15 'Das Sein in der Umwelt begenenden Seinden' (The Being of the Entities Encountered in the Environment). English edition: *Being and Time*, translated by John Macquarrie and Edward Robinson, Oxford, UK: Blackwell, 1962, pp. 91–102.

2

A Philosopher of Science's Opinion*

Gianluca Bocchi

CONTENTS

MEANING OF "PROJECT"

Question

To begin, I'd like to ask you the meaning of "project" from a philosopher of science's point of view, and to help us find some stimuli or suggestions for the people working in the management of projects, covering the role traditionally called "project manager."

---

* Transcription of an interview conducted by Francesco Varanini and Walter Ginevri on September 15, 2009.

Answer

From a philosophical point of view, the main point in a project is the bridging of two dimensions a philosopher would define as ontologically different; that is: a project is "the means to make true what is possible" or "an instrument to give an example of the future." In this perspective, a project manager's art is the highest. We know that in ancient times, bridge builders were called *pontifex* (Latin for "he who makes bridges," one of the names of the Catholic Pope). This role therefore covers a distinctly sacred aspect as well.

The human mind is immersed in an infinite ocean of possibilities, as we are well aware, so this infinity obviously needs to be made solid. We may then say that a project always includes a component of "reduction of complexity," in which reduction is not used with a negative meaning, but used in the sense of a choice and the risky responsibility of doing one thing instead of another.

Generally speaking, the human condition is very much based on reduction of complexity: just think of language, our fundamental means of reducing the complexity of the world by categories. So, the project manager in his domain continues doing what humans have always done in the course of history: create categories and refer them to the world. They are not always the best categories, but they work.

Similarly, projects possess both a lofty side and a pragmatic one: projects are not judged on whether they are true or false, measured with the yardstick of what is best, but rather upon the way they work, the way they make room for new possibilities for the project users.

So, what is the great problem we find in this book and in all our activities? The problem is that in the past people were convinced that the future was somehow contained in the present or even in the past, and in extreme visions even at the beginning of the universe. That is, as Galileo used to say, the future was thought to be just the object of the wide-ranging knowledge of a divinity or a demiurge. From Galileo until Laplace, this has been a regulating ideal: the supreme creator was the one in control of the passing of time.

In modern times the creator is conceived as the supreme planner of nature; any creator or planner is obviously not to be completely compared with the supreme one, Galileo used to say, except on a small scale, where this is possible. Therefore, even though the planner used to be a creator with a very limited sphere of creation, he tended to behave as a demiurge, a creator, dominating not only the project space, but also the time in which the project was taking place. There, this is a firmly rooted attitude in Western culture.

In words, projects have become totally laicized in a pragmatic manner. De facto, every project manager is faced with a very serious problem with this idea that the future is contained in the present time and wherever it emerges out of control the fault lies with us.

This is why it's interesting to understand that nowadays science, philosophy, and even, I dare say, theology and all sciences concerned with creation on any level (from the creation of the universe to the creation of an idea or a poetic text) are very much focused on the idea of "emerging." There is a lot of literature on this subject, but the fundamental idea is that some systems, the systems we call complex (the most interesting ones are life and human intelligence), are endowed with global properties that are not deducible from the sum of their parts. For instance, if we consider life, we may observe that there is nothing in the chemicals that create life cycles that is not physical/chemical. But life as a whole transcends the physical/chemical universe. In the same way, the human mind comes from a network of neurons; there's nothing intelligent in a single neuron, whereas intelligence is a systemic property. All this tells us the exact opposite of what used to be the classical vision, that the value of time in emergence is instrumental.

We have emergence in complex systems because there's a "critical time" in which the interactions between the parts of the system appear. In addition to time, scholars in complex systems tell us "diversity" is strategic too. If all component parts are not adequately dissimilar, global properties will be unlikely to appear. Today a very interesting literature about time as a creating dimension appeared. Should we consider Bergson beyond his nineteenth-century categories, his theory about time as a place of creation has been taken on some decades ago by key scientists such as Ilya Prigogine. The problem is, therefore, that the time dimension, far from confirming or debasing a project manager's work, should become part of the understanding and of the "fine" tools of projects. In short, we may say that a good project manager is the planner who finds confirmation in time. At worst, a good project manager is someone who doesn't want surprises in time. As the project is carried out, a good project manager would like his vision to be confirmed. However, a bad project manager faced with time interacting and proving him wrong, doesn't consider this an important acquisition, but rather a signal of his own inadequacy.

All this cosmology, considering time a source of deterioration and embarrassment is an old-fashioned cosmology going back many centuries. Science has advanced, but this is still many project managers' implicit epistemology in all domains. The idea is that the act of creating is not "inside" the process, but that it is an act coming before the execution of the project. The serious problem is, therefore, that a duality between project and time or between devising and execution will appear. In the use of words, as well, borders appear, or dualist epistemologies asking the brain to anticipate the world. However complex the brains of a team or a project manager, the infinitude of human brains is extremely low compared to the world's state of things. It's therefore a self-contradiction when, as happens in earnest in many domains, people think of anticipating all the possibilities underlying a project course in time.

Another of these myths is the myth of completeness. On this basis, we're faced with a very stimulating challenge, inasmuch as all the key words and procedures of a classic project have to be reconstructed with a different epistemological outline.

This doesn't mean not taking needs and requirements seriously. As an example, if we give up the idea of optimal results, and we think suboptimal or suboptimization, we may quote Francisco Varela [1], a neurophysiologist who studied the subject for a long time and coined a term I like a lot: "livableness." Projects must be "livable," and have to generate a seed for new possibilities: if a project is livable, its being an optimal project is not significant.

In such a context, my specific nature of philosopher of biology comes into play. Referring to optimal results, everyone should bear in mind Stuart Kaufmann's work [2] about the number of possible proteins. In biology as well we cultivate the theory that a living being is the result of an optimal project. It's funny how contemporary evolutionism was born at first by denying any outside violation, but nature later became quite similar to the "creator-demiurge." So nature is supposed to be optimal and we are supposed to possess the optimal biochemical composition by natural selection, and so on. Kaufmann calculates that a protein composed of 300 amino acids (a very simple one) may have an astronomical number of possibilities, more than the number of atoms in the universe, because there can be 20 amino acids in each position. In short, if nature tried one of these proteins every second, the history of universe would not be long enough to try them all. All this is to say that life works, but it's not the optimal result compared to given possibilities, because cosmic cycles would never suffice to try them all.

PROJECT MANAGER'S ROLE

Question

According to what you are saying, a project is only one of possible lives, or one of possible projects, so to say. But if a project is what it is whatever has presumed to come before, and the demiurge can do nothing about it, what is the project manager's role?

Answer

When we say that a project is one of possible projects, let's try and remember that the reduction of complexity is always a remarkable aspect. In short, I'd compare the project manager to a composer of poems, when poetry followed precise rules. Fundamentally, poetry was born as a very well regulated art, from the metric point of view. All classical epic poems were written with very rigid metric rules, and with these very rigid metric rules, when poetry was an oral art, very long poems were composed, real cosmologies in themselves. Because the poet was inhabited by art, the constraint and the rule were contained in the creative activity itself.

These very strong constraints gave birth to great creations. A project manager is something similar: The project manager takes very strong constraints seriously; he's faced with a very restricted angle of possible universes and doesn't roam the universe of possibilities. But he is still a very creative person because he becomes aware of the degree of freedom available inside these constraints. In other words, if the infinite state of things before constraints is so infinite that our mind drowns in it, the project manager, thanks to the constraints, has to do with a number of states of things that are infinite, but over which the human mind can prevail. Therefore, a project manager, like a builder of bridges, does not exactly realize the keeping in check of the infinite—as all mathematicians dream of doing—but the facing of the infinite number of possible worlds, much more humanely. All this occurs only if the project manager understands his own condition, that is, a simultaneous condition of craftsmanship and of creativeness. The project manager must consider himself an extremely creative and an extremely constrained person at the same time.

If either of those two elements is missing, the project manager is not a planner anymore, or, according to our criteria, not a good planner anymore because he could fail either for lack of a sense of reality or for lack of a sense of possibility. If a project manager doesn't understand that the strictest constraint still allows him an infinite number of possibilities, and that he has to face this risk of an infinite set, he is only an executor. On the other hand, a project manager who doesn't start from rules and constraints doesn't work on projects but on something else.

There are no projects without constraints at the start, and just as well. This applies to art, too, although contemporary art apparently accustomed us to the absence of constraints: the whole history of creative art (I was talking about literature) was born under very strong constraints. The contemporary debate is at first born from those constraints, too, then later extends them, denies them, and so on. Fortunately, a project manager stands in the same condition as a classic craftsman: constraints are there. The big mistake many project managers made, also in the last century, was to use the presence of strong constraints to infer a normal itinerary, in doing so nearly relieving the project manager himself of responsibility; the passage is from an artistic vision of the project, which is essential, to a mechanical vision of the project, as if the project manager were interpreting something written outside himself.

PLANNING TOOLS

Question

What tools can the project manager use, then? A traditional project manager gives a description of what he'll do in a project and afterwards checks the project. But if we think of a craftsperson project manager, what are the tools?

Answer

To give an answer, I use, after the artist, a second metaphor, the metaphor of the good scientist. What does the good scientist do? He questions nature. The experiment is a question to nature put in such a way that, whatever the result, whatever nature's answers, the answer contains the limit of the question. There's a teaching considered fundamental for one's research. The good scientist doesn't experiment banally, he tries experiments the result of which is really doubtful: whatever the result, he widened his knowledge of nature. Popper [3] met with success with his idea of falsificationism not because it's scientists' usual behavior, but because it's true that in science the falsification element is a relevant one. A kind of science in which you always only look for confirmation of your theories doesn't make sense. Therefore this component of a questioning nature is quite correct, inasmuch as you, who are an underset of nature, question a bigger set. So you already know from the beginning, as a scientist, that you cannot control nature and that the result of the experiment will be unforeseen.

There it is: let's think of a project manager who can enter this attitude that is not an attitude where any answer is possible. There's always a more or less ample expectation of the answers (for many of them you know there's a confirmation, for others a denial), until you get to the idea that you cannot be neutral regarding the response nature will give you because you, of course, read the response with your own instruments.

So we call a project manager, seen in the metaphor of the scientist accepting nature's response, someone who accepts the response of time, and more than nature's, that of the project's environmental and social context. In this respect, the project manager is even more constrained than the scientist, because the scientist in his laboratory can make everything add up. At the same time, the project manager needs a confirmation less than scientists did in the past, although the latter now do too, since science got involved with huge sums. As the project manager must conclude a journey that is constrained, obviously the replies from society or from the context do not diverge so much because the constraint is very strong. My intention is to underline the deep character of a project manager's activity. Although channeled between very thick walls in the construction of the quality in a project, the precise query, the precise denial, and the precise amazement are very important.

We usually think of a discovery as Eureka, Archimedes, and the change of paradigm. Actually there's a serious problem, not only in science, in the quality of our products. Even if ours is a routine job, our steps forward in knowledge can actually be radical. Fundamentally I call this dimension a quality dimension, as quality adds depth and insight to roughly defined domains that must afterwards take root or, as Varela [1] says, be embodied. The culture of a project is "constructive" in the respect that the project manager doesn't fundamentally discuss the major constraints, and his creativity stands within them.

Where then does the great problem in projects lie? In the fact that, many times, because the project manager knows he is under constraint, he doesn't consider questioning his environment every day is required. On the contrary, this questioning is necessary because, in the long range, the project manager may discover pathways that add quality, and may, for instance, allow one's project to build a network with other projects, or give a future to the project itself, or even help see one's project from different points of view.

All these elements introduce one's project in a wider context of symbols and cultures, therefore, from this point of view, two people with equal technical skill may be different as to their attitude. For instance, the more an architect keeps in mind the historical symbolic system of meanings in which his work is inserted, the more his work will be livable on the public's part. He may also neglect the context during the construction of the work: no problem will issue; the problems come later, because there's a lot of "not done," making the work unpopular or even receiving hostile reactions. Contemporary architecture provides us with a great many examples in these terms.

PROJECT MANAGER AS A DEMIURGE

Question

We could now start talking, from the project manager as heir of the role of god and demiurge, about the project manager, operating the project as the meeting place of different projects, of different people, of different viewpoints, where the observer influences the object of the research, and so on. If we widen our horizon, and we look at the project manager as just one of the subjects looking at the project, what do we see?

Answer

At this point we must go back and wonder who is a project manager, and this is like saying: who is a creator, who is a scientist, and so on. Now, without exaggerating, that happened in the last decades of the twentieth century, when the idea that there's no "subject" and that creation is impersonal, spread. Far be it from me to carry this idea to the extreme; there's no doubt that a successful project is an emergence in itself, and that the project's author is not a single person but a network of people and factors. If we then ascribe subjectivity only to human and personal factors, we could say that historical and environmental conditions are constraints in which the human communication networks are at stake. One could even go so far as to say that the constraints are actors as well. Now, without going into details we aren't interested in right now, one could say that, talking about human systems only, any project manager belonging to any organizational engineering or architectural structure actually is a "co-planner," together with many other actors going from the customers to the public, to the technical team, and so on. In this respect, Edgar Morin [4] always insisted that any book, any work of art, or any idea is always a collective emergence. Subjectivity is not excluded by this conception; on the contrary it gets greater room as subjectivity cannot create in empty space but needs a condition of communication.

This said, the project managers' choices come into play. There's a project manager who, for his very humility, considers himself not only an author but a co-author, and, understanding the importance of his own role, acts as coordinator of different actors. The project manager is not the exclusive author, but an actor managing other actors; therefore what may be lost in exclusivity is gained in awareness. However, the project manager who thinks of himself as the sole author creates a hostile relationship with the actors and the context of the play, and his project will be less endowed with emerging qualities as the result of limited interaction. This is similar to the opinion I often shared with Francesco Varanini [5] about writing: today there's a different way of embodying the fact of being an author or even just a creator of ideas. This is a generalization of the idea of "project manager" because one can go on thinking that an ego removed from others may exist and that ideas are a personal copyright (a banal and not so interesting attitude), and on the other hand that the author is always something collective. But saying that the result is collective and unpredictable doesn't mean saying that we're all on the same level. What the author of something, a book or of a work of art, always is, is the person assuming the need and the risk of connecting different points of views.

For this reason I think the metaphor of the *pontifex* (the builder of bridges) is quite suitable and probably ancient people had something similar in mind.

ABILITY OF CATCHING EMERGENCE

Question

To make it short and synthetic, we're facing the ability of catching emergence. What are the abilities and the soft skills we might figure out are needed to interact with what is emerging, to catch it in good time and to live the time?

Answer

The problem is that to define these skills we need to access some terms that are somehow worn out and others that we didn't study very well, such as listening and attention.

These terms seem vague, but they actually contain important psychological aspects. What do listening and attention mean? They mean that, assuming our control over what comes to us from the world is quite limited, and that at the start there always is an act of humility, the attempt is to widen the range of perception of these domains. Now, the problem is that these two terms assume a connotation of heaviness, as if listening or being attentive to the world requires an enormous ability of concentration. Actually the truth, in our opinion, is exactly the reverse. In fact, the attention to nature's cases on the scientists' part can be realized when a meaningful relaxation takes place and definitions other than concentration may be used.

So, the ability to listen materializes with a mixture of lightness and concentration. This was very clearly underlined by the late psychoanalyst Ignacio Matte Blanco [6], who said that, already well on in years, he used to play a game: when he was coming home in the dark, he tried to hit the keyhole with the keys on the first try. He realized that when he concentrated too much he failed and that when he was absent-minded he failed too. He was successful only with the right mixture of concentration and relaxation, which is what Eastern culture passed to us with the Zen culture. If in the East there's the Zen tradition, in the West we've the tradition of the *flâneur* (French for "stroller"), a person who explores an area without any precise direction, not only a town area, but a mental one as well: one may be a book *flâneur*. This describes a person who strolls in an area following a nonlinear route (even going back on his own steps) in a tangled topology but a still thick one, where the space is covered many times but from different points of view. Now, these attitudes, extremely precise from the point of view of cultural tradition, but not part of the training horizon of the project manager, are extremely solid. The project manager has to keep in his object's mental space, but he must have the ability to see it from different perspectives.

Therefore the project manager must move around the object. If we consider the project manager as an architect this is perfectly suitable because the architect should probably do what the *flâneur* does: where a new district is to be built he has to go through space and time around the place where the district to be designed will stand with this kind of attitude, a physical one as well. Not to live the space and time of the place, as many architects unfortunately do, doesn't make any sense.

Many athletes, the day before the competition, visit the Olympic stadium, not for a training session, but simply to go and see it. Now, all stadiums and platforms are the same, from a rational point of view. But what you get is this sense of marking one's territory that is a fundamental animal inheritance and triggers an archaic process of understanding. The art of a project is probably a very ancient art because, if we talk of *homo* (Latin for "man"), we mean *homo technologicus* (technological man). And if it's true that we have been applying technology for two million years and even Paleolithic technology is a technology needing a project, this means that there has been a long evolution in this attitude.

USE OF KNOWLEDGE

Question

Let's now add a passage. If we say that a project is an "emergence," something discovered or built, what is the use then of having prior knowledge, and what is the use of knowledge? Where does useful redundancy stand if every project is different from every other?

Answer

Knowledge is the knowledge of constraints. One must be careful not to flog a dead horse as there always is some component of constraint, rule, or standardization because the project manager is constantly acting in conditions of time and resource shortage. In the course of the history of art and architecture, some projects had very high resources on hand, but ours is not a slave economy anymore thus there's always a constraint (at least the constraint of the force of gravity, or the fact that one gets older and so on; we have inside constraints linked to one's mental abilities).

Therefore, understanding the nature of one's constraints is a fundamental condition. There are many constraints, and different ones, and so the more we find our place in regard to these constraints, the more, paradoxically, we understand the spaces for freedom: thus, more knowledge, more freedom. The paradox is that sometimes it is said it is better not to know too much, lest one freeze all action; in my opinion, in regard to the project manager this is not true. The project manager should know the condition of the constraints, in order to calculate precisely the perimeter of the space where he can operate.

STAKEHOLDERS AND CONSTRAINTS

Question

We speak of a project in relation to all those who are somehow involved, the so-called stakeholders, those bearing some interest in the project. These people have their own cultural, environmental, and historical assumptions too. Are these constraints as well?

Answer

Of course, and not only eliminatable ones. In fact, even if a patron with considerable resources were found, his requirements could be much more binding than what stakeholders with much more limited material resources have to face. If we look at history, the project manager/ customer relationship has always been a tormented one. Even projects with great material advantages had to face very serious constraint problems (creation in sixteenth century Rome or in the Medici's Florence was surely not easy). On the contrary, in environments comparatively modest from the material point of view, conflicts with stakeholders were fewer. In short, these constraints always exist: constraints of space and time and constraints coming from the final users, who are often not sufficiently taken into consideration by the project manager, who can be more solicitous in communicating with the main stakeholders. Some projects, therefore, from the project manager/stakeholders' relationship point of view may be considered a success rather than a failure in the long run.

Let's think, for instance, of projects for new cities such as Brasilia or the towns around Paris: those were wonderful projects from the point of view of collaboration between the project manager and political forces, but in the long run a lot of important things turned out not to have been considered.

END OF A PROJECT

Question

Then, as a paradox, when can we say that a project is finished?

Answer

Paradoxically, it's never finished. Many times, some projects have been qualified as projects in development. But even in projects that are considered finished, because the life of what we carried out begins right from the moment in which the project is finished, we can deduce that the project is never finished.

At most, there's a distancing on the project manager's part as if the project lived its own life by moving away from the project manager, by being "orphaned." But, if the project manager lets it go, the project is carried on by a community, we may say. Users' communities carry on the project abandoned by its creator. This subsequent life of projects can be as meaningful. Let's think of "reuse," a category that was considered marginal, and that became important in the transition from an industrial to a postindustrial society: we have so many things we don't know what to do with them.

Actually, any object of design or any architectural work is always the object of constant symbolic reuse. And reuse fundamentally means "reinterpretation." This is valid also for those particular project forms such as books and works of art.

As we know, after 2,000 years from the composition of some text or some work of art, the creation of that work is not finished even if its author has been dead for a very long time. Now, this doesn't mean that a project manager should always feel bound to his works. He can most probably create favorable conditions in order for the community to assume responsibility for the subsequent life of the project.

The great advantage of architecture is that it materializes all this. In the last decades, we have talked a lot about evolutionary materials: any building may have an interesting history, even if its function remains the same. Ancient people were better than we are in this respect. All Europe's churches have been remade in different styles. Darwin [7] himself wrote something interesting about the Duomo in Milan, saying that it's an evolutionary construction, being built for five centuries, and therefore its mixture of styles is not due to bad planning but simply to the fact that this is the way evolution works. This is the relationship between evolution and the idea of reuse.

Talking of reuse as if there were a fundamental use and then, if things go wrong, calling it "reuse," is wrong. All this is the rule in evolutionism. Now, because a project manager can never know, in the context in which he puts his work, which system his project will be part of, creating conditions that encourage evolutionary flexibility of the project is surely important.

PROJECT QUALITY

Question

This deals with what we consider the project outcome, or the project quality, and somehow its redundancy, necessary to create the conditions for its life. What do you think about it?

Answer

According to the object we are discussing, the possibility of a host of choices may be characteristic even if not so extreme: the choice of leaving parts of one's project incomplete, to be completed or to evolve in the future. We may therefore say that a project is never finished, although with some limits. Even the most perfected projects have a strip of transition, a fluid strip between "possible" and "real." A project is never realized all at once.

Let's think of the automobile or aircraft industries and of the idea of prototype to which they accustomed us. It's not only for commercial reasons;

this fluid prototype strip or testing is really needed in order to understand what possible evolutions were still unknown when the project was still part of future possibilities. In many domains the idea that when the project is half realized it can be reversed still stands.

This fluid strip may also have aesthetic values in itself. Let's not forget that prototypes made the history of design. In this respect, the story of the Jewish Museum Berlin is very interesting. There was a beginning phase in which the building was closed and it was used only from the outside, and, nevertheless, the outside was very strongly expressive. Later the museum, along with Libeskind's parts which made it famous, such as the Garden of Exile and the Holocaust Tower, was made available, but the museum's building was completely empty, without any collection. It was an extremely interesting case, where anyone could understand architectural volumes and aesthetic values without any misleading view due to collections or furniture. Then after two years the building was finally equipped and opened as a museum.

The transition phase hasn't been a preparatory space but a moment of great aesthetics, which gave way to something different in comparison with what it is now: as a result the people who lived through these three phases have a comprehensive understanding of the building. This is generally not done, but if one has to deal with this kind of dimension one understands a lot about the nature of projects and also because our tendency is to set the possible, the reign of reversibility, against the real, the reign of irreversibility.

A fundamental trend in the last decades has been the existence in projects of a somehow reversible part; in our world objects are transformed unpredictably more and more often. Even those who do not expressly subscribe to this epistemology of complexity acknowledge this need for a functional reuse. This is why today, between what is possible and what is real we find so many blind spots, or light spots, if we may call them. A lot of literature has been written about intermediate spaces as places of creation. Between the possible and real is an intermediate space into which one may retreat from reality. Obviously a project manager has to know how to move around in this territory. The most interesting things happen precisely after a project has taken shape. Some people underestimate this space as they are in a hurry to finish the project. Let's not forget that in the past the project manager and the tester were the same person: Porsche manufactured his car and then drove it, too.

REDUNDANCY VERSUS CONSTRAINT

Question

I'd like to understand the concept of redundancy better, inasmuch as it is considered by some as something to be conquered and negotiated in terms of further means and possibilities. We are saying here that redundancy needs first of all to be created. Is that so? And, if yes, how do you reconcile the concept of redundancy with the idea of constraint?

Answer

Some outer situation may obviously be more or less constrained, and therefore more or less redundant. But up to a certain point, these situations are relatively dependent on an inner redundancy that sits in the project manager's culture. "Culture" fundamentally is redundancy. It's the coexistence of several models, several languages, and several networks in space and time. This is what constitutes the difference between culture and specialism.

This is why Francesco Varanini [8] and I bet on the indispensability of cultural training underlying the productive system, especially when the latter wants to be connoted as innovation, because this is actually the sole instrument allowing people to move in contexts bearing strong outer constraints. Obviously, all this may not be sufficient: when the constraints are too strong, no culture holds up. At the same time, in contexts and situations that are different from the economic and temporal point of view, the resources of inner redundancy may become crucial. In short, which is the most common mistake on many project managers' and, generally speaking, many managers' part? To think these two kinds of constraints to be co-dependent, so that one answers great outer constraints with an even more constrained culture. The total opposite is true: if redundancy is missing on the outside, having it inside is a great component of security.

UNCERTAINTY AS A CONSTRAINT

Question

And if my outer constraints are great and uncertainty is a constraint in itself, I still have to find a way of discovering answers. Isn't that so?

Answer

Of course, and I would add that the outer constraint itself may be considered an object of experimentation. Under constraint, the spaces of possibility created by the subject make sense with the constraints themselves. The search for new materials was born precisely out of the need to overcome constraints that were considered insurmountable.

EXPERIENCE VERSUS KNOWLEDGE

Question

Another thing connected to this is the idea of "experience." Let's start from the goal of the project, usually a constraint given from the outside. After that, I gain experience; I use my experience, and so on. Now, are experience and knowledge the same thing?

Answer

In the term "knowledge" there's something consolidated and belonging to many subjects; in the term "experience" there's something emerging and totally subjective. This is why it's much harder to pass on experience than knowledge. The expert's ability is measured in this aspect. All this brings to light another problem, the problem of delegation, because obviously if in a project assigned to some team I need high skills, these cannot belong to every subject. This is where the project manager must be a leader, because he must decide every time where to delegate and where to experience something personally.

PROJECT AS A LIVING SYSTEM

Question

This takes us back to the "network" case as, in this sense, a project consists of letting this network of implicit knowledge emerge and be finalized to the goal. So, tell us something about the project as a "living system" where the brain belongs not only to the project manager, but all the stakeholders.

Answer

An interesting thing the neurological sciences tell us is that, because our mind is a network, this doesn't represent a great devolution: we just network our mind, a net, with other networks. Working in a team doesn't mean that my mind is networking with other people's minds. If my brain consists of many things and many modules, parts of my self are in constant relation with other parts of my self, and other parts might be very far from each other. Nowadays, we talk a lot about "mirror-neurons," and without going into details, the fact that some activations are very distinct, not in the brain as a whole but between some modules for some actions, means that the relation between myself and others is not only one thing; it is neither intense or less intense; sometimes we have a moment of stronger interaction with another animal.

STRUCTURAL COUPLING

Question

How does this structural coupling, this kind of "triggering," occur?

Answer

Probably because it's an evolutionary question. A fact still stands, as several experiments ascertained: for instance, if you put a ballet expert in front of ballet dancers and in front of *capoeira* dancers, some parts of his brain will obviously be more stimulated when he sees a ballet dancer than when he sees a *capoeira* dancer and vice versa.

This means that a kind of collective mind appears, with nothing mystic, but just made possible by association and experience, and this contributes to the emergence of collective behavior. Why is such a great space given to improvisation and gestures in training? Because improvisation in jazz music is anything but improvisation and it can be done only if you possess remarkable associations, the same for a soccer team. Both the football team and the jazz band are not banal projects, because both need a physiological time for these things to be triggered. What I mean is that probably the network represents the original condition of subjectivity. We keep thinking that the "I" comes first, and then comes the network. This assumption is wrong and so is the humanistic tradition telling us that we have the author on one side and the public on the other side.

For instance, the condition of archaic epics is totally different, because, until Homer lived or was invented, no one wondered about the authors of epic poems, because epic poems belonged to a whole culture. They evolved from one generation to another and the problem of who had started them lost its meaning. Even the subject matter of the Trojan War is not the beginning of the poem. The beginnings are the metrical structures.

In my opinion, we should revert to thinking that way, in the sense of thinking that whatever the project we are managing, the project has its own history, and we are not responsible for its beginning and for its end. We enter the story of the project while it's moving and we'll get out of it while it's still moving. I'm not saying all this in order to take the burden of responsibility off our shoulders, but to say we're important because we are unique and because at this particular moment some skills we possess are needed; the point is that instead of being demiurges we are good artisans, the same as nature is.

Evolutionism kept considering nature a head-down demiurge, even if an immanent, not a transcendent one. Only in the last decades has the idea that nature is an artisan, endowed with inner constraints very similar to ours, been taken seriously. These cultural assumptions are well rooted and imply extremely real consequences. For this reason nowadays the project manager and the organization must not wait for a revolution in the chief world systems, as quite often change goes the other way. Someone bases his implicit epistemology on what he does and then years later he realizes it isn't worth anything anymore. This "decentralization" makes every day's practice yet more interesting. If one tries to see how each of Libeskind's works is born, and how a district is built, one actually finds lots of symbology and intellectualisms, without any declaration as to the underlying philosophy. Philosophy is first embodied then theorized.

For this reason nowadays design is something culturally fundamental, going from the object of daily use to the car, because philosophy is first embodied and then declared, or not even that. If we waited for a philosophical revolution to come first, and then for a revolution in human practice, everything would be lost: philosophy would be like the owl getting up at dawn. The fact still stands that, even if these practices are well spread, on the level of cultural transfer and academic and professional training neither the most interesting cultural and scientific results nor today's actual practice are taken into account.

This is why the testimonial is important: it's a way of learning experiences that as such either cannot be theorized or, worse yet, when they are theorized their communication conveys the exact opposite meaning to the original. Therefore this gives us a training problem. For the project manager, who is operating between the real and the possible, an interdisciplinary training sphere becomes essential. All this gives technique a new role as well, as the retrieval of handicraft can be set at the intersection between humanistic and scientific knowledge.

Now we know that the art of the project, technique and design, are not inferred from science but need science. Therefore we can abstractly state that we are free from science; on the other hand it's totally different to understand that there's a quite definite artistic component, in the precise sense of this term. The retrieval of craftsmanship as a kind of knowledge is quite distinct, inasmuch as a craftsman is a person possessing "singularity." Therefore he is perfectly in tune with experience as an individual component. Let's go back to the previous point, that knowledge can be shared whereas experience only rarely so and it can be communicated only after it's already crystallized. Why is the "master" important? Because he is the one involving his pupils in the experience "here and now." The teacher doesn't.

PROJECT MANAGER'S TRAINING

Question

Does this mean that one learns to be a project manager only if one works alongside another project manager?

Answer

Certainly. This because the project manager is found in a context: he lives in the "here and now," All these things emerge when one speaks of complexity but the problem is that one doesn't start all over again every time. There are actually many seeds, many practices and considerations that we are trying to render systematic. What I mean is that many times this way of speaking gives an exaggerated idea of discontinuity compared to reality.

It still means that, because in this strange reality some cultural attitudes don't coalesce, the fact of connecting things becomes the true discontinuity.

If we weigh things taken singularly, such as today's statements, we see they are intelligible, if not even common sense. In a more co-ordinated picture, such as in books and in our activities, they are still worth something. This represents the cultural matter of complexity. Complexity is often criticized because its statements are truisms or banalities. This is somehow true if we think that language itself is the first problem of reduction of complexity; philosophy's only goal is making the complexity of the world livable.

However, one must understand the meaning of this kind of reflection in an age where undoubtedly the complexity of the material world has rhythms, times, and spaces totally different both from the ancient world, and from the modern one too. Our working on it, with the awareness of being craftsmen is essential. Our cultural and interdisciplinary project existed before us, but this doesn't mean we shouldn't do our utmost in this direction.

REFERENCES

1. Varela, F., Thompson, E. and Rosch, E. (1991). *The Embodied Mind: Cognitive Science and Human Experience*. Cambridge, MA: MIT Press.
2. Kauffman, S. A. (1995). *At Home in the Universe: The Search for Laws of Self-Organization and Complexity*. Oxford: Oxford University Press.
3. Popper, K. R. (1959). *The Logic of Scientific Discovery*. New York: Basic Books, Chapter IV.
4. Morin, E. (2008). *On Complexity. Advances in Systems Theory, Complexity, and the Human Sciences*. New York: Hampton Press.
5. Varanini, F. (2003). *'La restituzione poetica' in L'irresistibile ascesa del direttore marketing cresciuto alla scuola del largo consumo*. Milan: Guerini e Associati.
6. Rayner, E. (1995). *Unconscious Logic: An Introduction to Matte Blanco's Bi-Logic and Its Uses*. New York: Routledge.
7. Quoted in Gould S. J. (2002). *The Structure of Evolutionary Theory*. Cambridge, MA: The Belknap Press of Harvard University Press, page 19 and following.
8. Bocchi, G. and Varanini, F. (2004) *'La macchina analogica', in Boldizzoni D. and Nacamulli R. Oltre l'aula. Strategie di formazione nell'economia della conoscenza*. Milan: Apogeo.

3

Testimonies and Complexity

Fernando Giancotti

CONTENTS

WE ARE WITNESSING

I am actually witnessing a fascinating event. Less than one year ago I was invited, together with Francesco Varanini, Gianluca Bocchi, Alberto De Toni, and Luca Comello, to a convention called "Projects and Complexity" to speak about leadership in complex environments. Later I was visited by two "Complexnauts" and addressed a lot of questions. Only few months ago, in May, I shared with 25 project managers a short and intense seminar called "Leadership, Complexity and Projects: New Ideas for New Approaches."

I remember well the excitement of those events, of the discovery of new perspectives, of the doubts, of the vivid intellectual curiosity, of the ensuing discussions. From the Discovery to the proposal a very short time elapsed, because the Proposal of the Complexnauts shows a deep comprehension of the Discovery, but goes much further, and translates it into concept and operative indications that can renew a way of thinking and managing the work by projects, and, I daresay, work in general terms. What kind of phenomenon is this? What shape can it take and what can its contents be? What will its effects be? Let's discuss it, at first maybe accepting Bice Dellarciprete's and Andrea Pinnola's suggestion: let's tell the story.

LET'S TELL THE (RE)DISCOVERY

The first useful story is the history of the Discovery. I observed it through single events, each happening a few months after the other, and I cannot give details. But it appears similar to mine, qualitywise, and similar to many other people's who changed their *Weltanschauung* (way of looking at the world), and now look at the world as a complex system.

As a kid I saw the "conquest" of the moon; science astonished me. I grew up with a strong rationalist imprinting, I studied engineering, I worked in a highly technological world, the aeronautical one, and I received "traditional" military training, suitable for the Cold War, mostly a slow and evolutionary phenomenon.

Theory, not only in the military world, proclaimed: leaders give orders; subordinates carry them out. I was also trained in critical thought, and history as well, by some good classical studies teachers in high school, and while reading about evolution, ecology, and some evocative quotations by a Native American chief or from the Eastern thought, I opened my mind to the doubt that something didn't add up, that some dark matter was present around the linear relations of cause and effect that fascinated us so much, something I couldn't see but the effects of which found their expression in a powerful and uncontrollable way.

Nevertheless, it's above all the work and life experiences that confirmed those doubts and organized conceptual and operative answers, together with other readings, most about leadership. A long journey, together with research opportunities introduced me to the new science of complexity and chaos. And this science, although taking the first steps, elucidated to me a lot of what I was seeing and changed my way of looking at the world: from the way I face change in a project, I pursue an aim, I bring my children up, I look at my career perspectives, up to how I take a holiday and much more.

The Complexnauts, and I write it without quotation marks because they deserve it, understood in a few months a not at all foreseen, and often quite long, journey applying the same courage, commitment, and generosity of explorers. They heard of the existence of the Land to Be Found from Francesco Varanini, were led and assisted by Carlo Notari's and Walter Ginevri's leadership, were told stories and given indications by many who made the journey before them, the expert navigators, who were called to contribute.

But to them belongs the deep understanding, to them belong the difficult exploration, the discovery in that continent, that is all continents, a new land potentially able to give new answers to actual problems. I think they also acquired another way of looking at the world. It's a fascinating story, the implications of which we discuss again. But it's an old story. It's only that we had forgotten it. Our hunter–gatherer ancestors, and the people of cultures who survived long enough to be registered by history, were an integral part of the world they were living in, of the "Whole."

The Native American chief Seattle wrote to the American President Pierce in 1855: "This much we know: earth doesn't belong to man, but man belongs to earth. Man didn't weave the net of life, he's just one of its threads. Everything he does to harm this net, he does to himself. What happens to earth, will happen to earth's children too. We know: everything is linked, as blood links the members of one family."

For centuries our farmers too tilled the land, built their tools and everything they needed, hunted, and lived in a world that was known as much as necessary and possible, but known as one and unitary. Humanism, putting man at the center of everything and creating many disciplines in order to penetrate deeper into the sphere of interest of intellectual exploration, designed the ideal of a many-sided man, interdisciplinary as to activities, able to look at the world from several artistic and philosophical perspectives, whose symbol is Leonardo da Vinci, with many other examples. Eastern thought, that interested our society so much, although being much more diverse than we usually think, embraces the idea that "the Whole is one." In the military schools of the new globalized world, an important example of the approach to complexity is Sun Tzu's famous *Art of War*, implying a deep systemic thinking. This unity, even with all the limits and specifications that can be singled out beyond myth by a serious historical approach, has been shattered, especially with the extraordinary production of wealth, and the complexity linked to it generated by the industrial revolution.

In a book close to this research one can read:

> Since we human beings are building this complexity [...] we should know how to manage it. Up to now, we didn't do a very good job. The industrial culture's answer to this complexity was to simplify through specialization

and through compartmentalizing knowledge. Impressive advances were achieved by the simple focusing power that specialization implies. But too often this approach led to a fascination that kept us from seeing the forest for the trees. We missed many of the unintended consequences and failed to think through solutions for the problems we were creating.

Thus, technological advance in specific fields has meant devastating environmental impact elsewhere. Taylor's "scientific management" of work meant widespread alienation.

Organizational "rationalizations" resulted in huge centralized structures, where individuals felt little responsibility or sense of belonging. And value systems, the very glue of society, seem to be in a permanently unstable balance, as diffused phenomena such as drugs, crime, violence, and the breakdown of the family too often remind us [1].

The specialized cognitive approach has been at the same time an effect and a cause of the wealth produced, and of the extreme articulation of our systems. It often didn't work well, but in the Western world created a very powerful mode of the collective action. Nevertheless, today its troubles and problems are amplified by globalization and by the different geopolitical picture, and greatly questioned by the extraordinary revolution in information: the fragmentation of our thinking shows its limits all the more often and all the more obviously. Meanwhile, our thinking began to look for cognitive models that, with the expected difficulties and resistance, could represent an alternative to the "mechanistic" view of the world and to its obvious shortcomings; we may say, we begin to look for the lost "comprehensive view." And the people applying themselves to this research sooner or later meet the science of complexity and chaos, offspring of considerations coming from traditional scientific disciplines, and still aimed at identifying ideas able to explain complexity, the whole we are part of, the relations out of which "everything is bound together." Here we are back where we started, when the world was much simpler, and one: a very important rediscovery.

But, as the dynamics of complex systems show us, a journey, a set of changes, can never re-create what has been in the past as there is a strong "dependence on initial condition" and the flapping of a butterfly's wing is enough to change the course of events forever. In the meantime, using our linear logic we have been to the moon and we know how to do extraordinary things. It's a matter of using this logic in a new way, applying a deeper and more extended way of thinking, systematically trying to see everything as a whole, and to be able to understand the system, the essence of its relations, and to learn how to operate in uncertain and unstable environments. Finally, the further quantum leap in the level of complexity we live in compels us to invest in all this, but gives us new cultural instruments as well, and machines able at last to integrate knowledge instead of breaking it up, able to connect it instead of closing it in compartments, and able to build complex knowledge systems, the implications of which have a fascinating heuristic value.

The (re)discovery reproposes the ancient perception of the wholeness of things, but in a more powerful way, because it's not an alternative to what linear thinking, extraordinarily effective in each domain, implies: it integrates it, it doesn't deny it, it widens it, and goes on, without considering unlikely restorations. Furthermore, it accepts its own mistakes, because its intrinsic dialectic is apt at correcting them. In short, it welcomes old skills in a new vision, able to widen their limits.

All this is reflected in the Complexnauts' work and in the implicit comforting wisdom with which they explore the new world in order to actually improve the world in which they are operating.

THE (RE)DISCOVERY AND THE BUTTERFLY

There's another perspective that can be used to look at what is happening which I think can be appropriately added to our story. According to what I have been able to observe, the rediscovery is a complex event, happening through the very mechanisms it is describing. The reflection about the problems project managers are faced with in the development of their carefully planned projects, is probably widespread in the professional community. In this cultural system, shaped by common rules and inner communication fluxes, a particular kind of dynamics has been triggered in a specific domain: the community of project managers referring to PMI Northern Italy Chapter. Reflection over these problems was touched off by an event happening by chance: the opportunity, under Francesco Varanini's proposal, of examining it again in the light of complexity theories. This opportunity could have been dropped. But for a series of factors, the system of leadership relations and the "ethics" in that environment opened the way to the exploration of new perspectives, taking the risk, decentralizing the way of working and co-ordinating it through a shared vision. In the project managers' community, and particularly in that small environment, an extraordinary hypercycle, a set of mutually reinforcing feedback loops, started.

The quantity and quality of exchanges multiplied regarding the shapes of knowledge and of operations in the community itself mulitplied, up to the quantum leap, a catastrophic bifurcation that transformed this hypercycle, the discovery, in proposal. Translating the language of complex systems into that of human systems, leadership, ethics, and human capital of the above-mentioned small domain have elaborated a new conceptual and operative paradigm: a possible revolution in organizational practice.

This other short story told under the key of complexity doesn't represent only a meta-narrative about the discovery of complexity by a complex system, or by a part of it. It is above all an attempt at presuming what will happen later, projecting what happened in a possible development of the behavior of the system as a whole. In regard to this, I have been especially careful in connoting my words as possibility when talking of the effect of the proposal. This is because there's no guarantee that the hypercycle started by the Complexnauts can open a wider hypercycle, up to the production of a generalized change for the better in the organizational world, that could be measured, for instance, through a drastic revision of standard best practices. But, if there are no guarantees, there definitely is an extraordinary opportunity.

The basic subject, the relation between the planning functions of our complicated organizations and the management of reality, is shared by all relevant environments, by the corporate world, by the organizations in general, by the military world, and by governments themselves, all struggling with the most complex human systems, as they consist of a great number of systems. Project managers all over the world share a base and a language and are used to comparing experiences, and this helps the basic function of communication. This proposal may possibly, may even likely, provoke a discussion, as well as predictable resistance, and this in its turn may trigger a further wider hypercycle, a renovation movement. Further on, project managers are across a great number of organizations and are growing in number and influence. They can be an important cultural contamination vehicle in our world, the possible spreaders of a way of knowing and operating that suits what is happening today in the world to a greater extent than what we have seen at work up to now.

Western thought, but I would say, just human thought, needs growing up. The narrow mind [2, p. 187] we are endowed with needs investments in its ability to comprehend and act in the world it has created, and so often mismanages. The proposal is an attempt in this sense and there are conditions today allowing a system's reply. A strategy able to influence this system and to enhance existing conditions is therefore required. The flapping of butterfly's wings in Milan may cause a storm in Pennsylvania. They are worth flapping, those wings.

SEAMANSHIP, AIRMANSHIP, AND LEADERSHIP

But butterflies had been flapping their wings for a very long time before we came down from the trees. Our forefathers immediately had to face the complexity of the environment, using apt strategies well before the industrial revolution. Let's see just a few of them.

Seamanship

In his contribution to this book (Chapter 1), Francesco Varanini spoke about Captain FitzRoy, with his *HMS Beagle*, and Captain McWhirr with the steamer *Nan Shan*. Above all, he told us about their navigation in the unpredictable seas of their time, and their way of conducting it: keeping the goal of the journey well in mind, but looking beyond at the same time; trying to find the technical instruments and the handbooks in order to reduce uncertainty and staying well aware that only the judgment of the commander may integrate the whole of the situation of the moment, unique as to place and time, with the ability of feeling, of mentally seeing, of choosing what to do and when, and of facing the unexpected, because a good captain is able to face the tempest, not only to steer his ship in good times.

In Chapter 9, "Leadership and Complexity," Stefano Morpurgo uses the metaphor of the sailboat to show the perspective needed to steer an unpredictable system. The skipper has to know the winds, but can never influence them and sometimes not even forecast them. He may have to face dead calms and tempests, but he will have to steer his boat to its destination all the same. He will have to prepare the crew, but nevertheless he'll test their actual value only in an emergency.

The abilities of a good sailor are summarized in the expression "seamanship" (in Italian "seafaring art"), which dictionaries define as "the art of operating a ship or a boat." The very term "art" describes by itself how technical instruments, handbooks, and codified proceedings are not enough for real seamanship. This art, the knowledge and wisdom it implies, has been and is the strategy to face the sea, the winds, and the many dangers, minimizing risks and bringing the ship to its destination, or at least a safe haven, in an environment we cannot control, possibly even with higher utilities than forecast in the first place.

Airmanship

But this approach did not become outdated with steam navigation or restricted to the surface of the sea. Nowadays, as in the past, the most up-to-date aircraft require airmanship to operate effectively and safely, interestingly redefined. Some time ago I ran into a book, *Redefining Airmanship* [3], where one can read: "It suggests a system thinking approach, in which each element of airmanship is seen as making an impact on the whole, in a dynamic and complex human equation." And: "Breakdowns of airmanship are most often caused by failures of integration and not by any lack of skill or proficiency." In order to explain the problem, the author, an experienced pilot, quotes Peter Senge [4, p. 3]:

> From a very early age, we are taught to break apart problems, to fragment the world. This apparently makes complex tasks and subjects more manageable, but we pay a hidden, enormous price. We can no longer see the consequences of our actions; we lose our intrinsic sense of connection to a larger whole [...] we try to reassemble the fragments in our minds, to list and organize all the pieces, but [...] the task is futile—similar to trying to reassemble the fragments of a broken mirror to see a true reflection.

I have been flying for years and still do, even if not much. I had to start all over again several times, as I changed aircraft, operational specialty, or type of activity, or just to retrieve the practice I had lost. But the approach is the same, regardless of our skill at a given moment. There is a professional culture urging you to study your aircraft well, to know the systems and the emergency handbooks, always to prepare the mission before the flight and to examine it afterwards, ready, or rather willing to find one's mistakes to be able to set them right, using the necessary humility.

It's a culture that makes you systematically evaluate risks and never accept unacceptable ones. It's a culture that considers self-discipline a value, in order to do what you know is right without dangerous shortcuts on one hand, and in order to accept the operational risks that belong to you. It's a culture capable of producing sayings such as, "A superior pilot uses his superior judgment to avoid using his superior abilities"; "There are undisciplined pilots and old pilots; there are no undisciplined old pilots"; "In air battles the winner is who makes fewer mistakes." And it's a culture that sets the Situation Awareness, the comprehensive awareness of the situation, of the whole picture, and the Judgment, our best interpretation thereof, as cognitive fundamentals of one's activities. Significantly, in the rigorous flight handbooks, where the actions to be enforced in specific emergencies are recommended, this warning can be read: "These procedures detail actions to be carried out according to the knowledge of the systems aboard and of experience. Nevertheless what is indicated must always be applied to the contingent situation and can in no way replace solid judgment." This culture produces airmanship, knowledge, ability, and wisdom in order to interact with the complex system of environment, machines, organizations, and men operating inside them, mitigating risks and sowing opportunities in order to reap the intended results, and when possible, also unexpected, still useful ones.

Finally, airmanship is not a beautiful theory: it was born out of the mistakes made and of the corrections found. A very complex system shaped it. This word emerges from the blood of many airmen.

Leadership

English words bearing the suffix "–ship" generally belong to two categories: those indicating a condition or a belonging, such as citizenship, hardship, membership, scholarship, and those indicating a set of skills or knowledge, with a connotation of "applied art," as the ones to which we referred.

Leadership stands together with seamanship and airmanship. Leadership implies knowledge, ability, experience, and wisdom about the collective action and the environment, the world where it takes place, in order to interact successfully with its complexity. Its mechanisms are extremely ancient, born as they were out of the need, shared by all social animals, to balance the interest of the single member with that of the group, and the group's with the ecosystem, while ensuring the competitive advantage of co-operation and therefore survival. On that account leadership has always been a complex phenomenon, shaped by the environment, and by the unrelenting process of natural selection as well, at the cost of countless deaths. The small groups of hunter–gatherers in which humanity lived for 99.95% of its more relevant evolutionary history developed the social modes we carry in our genes this way. This "flat" leadership relationship, co-operative and functional, that evolved for hunting, is our natural one, the suitable relationship for the natural environment in which man was living in perfect integration, although in much peril.

The agricultural revolution, with its division of labor, the accumulation of wealth, the social stratification, and the numerical multiplication of human population made the natural model of leadership inadequate, as it also had to be exercised over much bigger domains, over faraway and unknown people. The great number of cultures born at that time gave different answers to the new problem: how to make the mechanisms of collective action work in big social agglomerations and complex organizations as well. Many answers clearly showed the fierceness of the superpredator, man, when beyond the control of the tight personal relations of the small group.

The complex nature of social and organizational dynamics made its first sudden qualitative leap by setting up political institutions, different kinds of organizations, wider economies, and contacts between cultures. The lapse of time for change, however, can be counted over many generations. If change indicators are linked to the phases of humanity's life, one can realize how slowly things changed over millions of years, and still did in the historical age. Although living in a complex system, people generally saw very little change over their lifetimes. The industrial revolution altered all this. In the last 150 years everything changed. The wealth produced increased enormously, although very unevenly. Social systems, economies, communication, and organizations became much more complex. The system made another big quantum leap. We are living today in the information revolution, which started a few years ago with an unlimited multiplication of data, news, and knowledge exchanges. If we go back to the change indicators we mentioned and we check their history, we see that the speed of change increased beyond belief. Our problem lies not only in interacting with a complex system, which became much more complex. It doesn't even lie anymore in trying to keep up with change. It rather lies in living and acting in extremely fast change, which, furthermore, keeps speeding up all the time. Our world's complex system is creating quicker and quicker hypercycles. Keeping up doesn't make sense, is not enough anymore. We must live through frequent "catastrophic bifurcations" and really have to know how to move "on the edge of chaos."

What about Leadership?

Today, as many years ago, leadership is the strategic answer we may give to the problem of our existence as a species, but this time in a new, very fast world. It's the strategic answer we can give to the problem of organization and project management in that world. In order for it to suit the quick change and the new complexity, ours has to be an answer that doesn't imply abstract *noumenons* to lay on top of reality, but rather aims at understanding the environment and at perceiving trends and vectors. It has to invent the future and avoid the pretension of planning it in a deterministic way; it must continuously sow opportunities; it must start and always keep the initiative, but be ready to address it elsewhere if the situation and our judgment suggest otherwise, ready to react to unexpected threats but much more to catch emerging opportunities, wherever they may come from, looking for them especially in trouble and mistakes; it must have good judgment in assigning priorities and choosing what to do and when; it must be effectively communicative and captivating and to seek questioning and discussion all the time, to direct change as required by the future we want to build.

But today, as in our ancestral yesterday, this leadership doesn't mean just a single chief's leadership, but a distributed leadership of which the chief is the promoter and the warrantor, where wisdom is shared and increased by those who can see from different perspectives and on wider horizons. We need a leadership able to guarantee the "ethics" through which to look at the world and to let "simplicity beyond complexity" emerge: making everyone understand the essence of the event we are sharing. There's no seamanship on a ship if this lies only with the commander. The crew is not safe if the only airmanship is the pilot's. Lastly, where seamanship, airmanship, and leadership are present, no one considers himself the ultimate expert. We already know that navigation and sea, sky, and flight, our people's guidance and the world where through this guidance we act, will keep surprising us many more times. What we can say to ourselves is precisely: "We are always ready for anything," ready to operate in the unexpected, definitely, today more than ever.

SO WHAT?

I think that all we discussed so far brings us to a simple conclusion, "beyond the complexity" of the analysis. To make it short, I can quote Francesco Varanini when he says in this same book that a project manager shouldn't squeeze our valuable projects onto a plane and represent them one-dimensionally and superficially. We should be able to read in them a future beyond the goals we set, in order to catch all emerging opportunities. The project manager must take charge of the complexity he operates in and be able to live the *kairós* beyond the *chrónos* that is imposed on us from outside. An ability of feeling, of mentally seeing should be exercised in order to choose the right thing to do, at the right moment. Ethics should be used to look at the world, because it is a fundamental frame of reference. And wisdom, which is more than reason, should be applied. Wisdom is also moderation, balance, and knowledge through experience.

But if all this is true, and it is true that if "management is doing things right, leadership is doing the right thing," [5, p. 18] what ensued from the debate, the (re)discovery of the Complexnauts, is a rediscovery of leadership as well. A leadership born in faraway times, in order to link up the complex system of the environment where we live, and that still preserves its basic qualities, but requires new cultural and cognitive instruments in order to face today's world's complexity and speed. The works that follow attempt both things, looking for a model "agile in complexity." And if semantics means substance, maybe the name for the new figures outlined, from whom much more is asked, is not "project manager" anymore, but plainly, "project leader."

REFERENCES

1. Giancotti, F. and Shaharabani, Y. (2008). *Leadership agile nella complessità. Organizzazioni, stormi da combattimento [Agile Leadership in Complexity. Organizations, Combat Wings]*. Milan: Guerini, p. 35.
2. Giancotti, F. (2001). Strategic leadership and the narrow mind: What we don't do well and why. In *AU-24, Concepts for Air Force Leadership*. Maxwell AFB, AL: Air University, p. 187.
3. Kern, T. (1997). *Redefining Airmanship*. New York: McGraw-Hill, pp. 7–25.
4. Senge, P. (1990). *The Fifth Discipline: The Art and Practice of the Learning Organization*. New York: Doubleday, p. 3.
5. Lester, R. I., and Kunich, J. C. (1997). Leadership and management: The quality quadrants, *Journal of Leadership Studies*, 4 (4):17–31.

4

The Shared Vision as a Change Engine

Alberto Felice De Toni

CONTENTS

PROJECT MANAGEMENT: ORIGINS AND DOMAIN

Project management is one of the central lines of managerial studies. In the beginning of the 1960s this discipline already appeared unquestionably important in the sphere of corporate culture and was apparently heralding, as was later confirmed, considerable changes in the promotion and management of innovation processes.

However, originally the approach was distinctly technical, the organizational aspects being somehow overlooked and the methodology restricted to specific domains, such as projects for large works (in civil engineering) and projects for the development of new products in industrial and information engineering. Moreover, project management was thought to concern only project or program units.

Subsequent developments changed the contents and the scope of this discipline, and came to encompass any project typology, including projects for strategic, organizational, and managerial change, from the point of view of an integrated company's management where all domains contribute to the success of projects and projects are implemented in all domains.

Frederick Taylor and Henry Gantt (who can, respectively, be considered father to scientific management and to project management) were both engineers, born at the beginning of the twentieth century. This was not fortuitous. Management was born inside Ford-style factories and tested on production organization, today known as operations management. From operations, project management kept developing, throughout any corporate domain. Because project management was born in factories this explains why project managers are required to have both technical–scientific skills (i.e., in basic technological science and plant engineering) and economical–managerial skills. Systemic, quantitative, and design and construction of models approaches on one hand, and planning and implementing skills on the other hand, are considered as important as the above-mentioned capabilities.

Planning and implementing skills, as well, require a target-orientation, a problem-solving approach, a methodological rigor, and a disposition to measuring. Characteristics pertaining to co-ordinators of interfunctional groups are also required: leadership, attitude to change, learning capacity, communication, integration, and process management. A project manager's critical success factor lies in his ability to interpret environmental change with sensitivity and creativity and to manage the problems concerning the subsequent changes in the organization.

PROJECT MANAGEMENT AS MANAGEMENT OF BECOMING

Production management evolved deeply in the last century: from the Taylor–Ford model of mass production to the Honda–Toyota model of Lean production. The management of innovation projects still does not seem to express its full potential because it is too often cut off in a functional organization context, with all ensuing communication, co-ordination, and integration problems. In the last few years the dismantling of this context started through "process orientation," now the fertile breeding ground for a change for the better in the application of project management to the organization as a whole.

Operations management and project management represent, for the company and the corporate systems, the management disciplines of, respectively, the steady condition and the transitory situation. Competition runs faster and faster and is pushed by accelerators such as technological innovation and market globalization. The latter forces companies and their supply and distribution networks to unceasingly redefine products, services, markets, production and distribution connections, production technologies, information and communication systems, organization structures, and managing processes more and more frequently, to the point that temporariness acquires a steadiness character: steadiness doesn't exist any longer or becomes less and less important; on the contrary, temporariness becomes the actual functioning modality of companies, engaged in never-ending changes.

We are, after all, rediscovering in the managing sphere what the pre-Socratic Heraclitus sensed a few thousand years ago in relation to the universal flow of things: *panta rei*, everything flows. The dispute between being and becoming takes root in an old philosophical case going back to the very origin of western philosophy. Parmenides maintained that multiplicity and change in the physical world are illusory and asserted that existence is real: unchanging, eternal, and indestructible. According to Parmenides, the philosopher of unity and identity of existence, change is an illusion, a blunder, and everything is fundamentally immutable.

Heraclitus, one of Parmenides' contemporaries, opposes the latter's thought. He, on the contrary, can be considered the philosopher of change and becoming: "No man can bathe twice in the same river, because neither man nor the water in the river are the same." The whole world is considered as an enormous eternal flow, where nothing is ever the same, because everything changes and undergoes constant evolution. For these reasons, Heraclitus identifies the shape of existence in becoming, as everything is subject to time and evolves constantly. Further on, he maintains that change and movement only are existing and the identity of immutable things is illusory: according to Heraclitus, everything flows.

Parmenides' logic of being won over Heraclitus' logic of becoming thanks to Aristotle's metaphysics; therefore philosophy first, then science, was founded on what became par excellence classic logic. According to Aristotle, becoming, an everyday experience, is but the passage from one kind of being to another one. In short, being is the only reality and becoming is just one of the ways of being. Aristotle formulates the ideas of potentiality and actuality. Potentiality generally represents the possibility something may change or assume some particular "shape." The act is the realization of that change, and represents the actual product obtained as a result of the change. For instance, a chick potentially is a cock, the same as the cock is the chick put into action. According to Aristotle, the act is superior to potentiality, inasmuch as it's the cause, the meaning, and the aim of everything that potentially exists.

The logic of becoming ceased to exist for over 2,000 years. The comeback to Heraclitus was proposed by Hegel, with his dialectics with which he explained the dynamism of reality through thesis (being), antithesis (nothingness), and synthesis (becoming). According to Hegel, the being is the beginning; it doesn't need any other concept at the source. It's the most undetermined concept of all, because any determination implies a relation/opposition with other concepts. Nothingness, although representing the antithesis of being, the utmost of indetermination, finally identifies with it, and synthesis, or becoming, is therefore generated. The truth of being, as that of nothingness, is their unity, and this unity is the becoming. The unity of being and nothingness is not a complete leveling, it's diversity, and this identity makes us realize the contradiction in reality.

In explicit controversy with Parmenides, but with Aristotle too, Hegel maintains that becoming has supremacy over being. Whereas according to Parmenides and Aristotle the being cannot not-be, it's noncontradictory; it's one. That is to say, everything is identical to itself. According to Hegel the being is and is not, it is contradictory, and divided into dialectic polarities that contradict and synthesize each other. Everything recalls its opposite, and is at the same time itself, its opposite, and the synthesis of both. What ensues, for instance, is that "falsehood is but a moment of truth."

If we may ascribe operations management to the philosophy of being, we may as well place project management in the wake of the philosophy of becoming. In the final analysis the basic thesis is that, as everything becomes, then management science and art do not concern "steadiness" but "transitoriness." What follows is that, in order to manage continuous organizational change, the typical project management approach is fundamental, and that is exactly the approach pertaining to "becoming." In overall synthesis, "management" is always "management of becoming" and the future of management is found in project management as the set of principles, methods, techniques, and tools apt at managing change.

THE DRIVING FORCE IN CHANGE: DREAM, VISION, AND MYTH

If management is always "management of becoming," a key question has to be asked. What are the driving forces in change? In order for this question to be answered, we have to make a distinction among persons, organizations, and society, that is dreams, vision, and myth. But let's proceed with order. We follow a pathway that brings us to state that the true driving forces in change are dreams as the individual's imagination, vision as the organization's imagination, and myth as society's imagination. For a project manager, in charge of a company, being aware of dreams, vision, and myth is as fundamental as for a racing driver being acquainted with the car's engine.

Dream as a Creative Source

"Probably no human experience is as distinctly creative as dream. No phenomenon is more endowed with unpredictable transformation potential. No moment is more inventively poetic, that is—in the etymologic sense—full of *poiesis*. The unforeseen, the illogic, the unthinkable, the unnatural become, in dream, most naturally obvious" [1,2].

Dreams reveal the environment we live in and connect us to it. Dreams take us to the region of knowing/not knowing, the seat of truly creative thought, and where a pragmatic approach to reality can still be undertaken. Dreams "think" reality. Dreams are the mental space where known and unknown co-exist, and where the unknown has neither shape nor time, or rather stands beyond and ahead of the concepts of time and space on which a great part of our mental functioning is based and to which it is confined [3].

"I have a dream: that my four little children will one day live in a nation where they will not be judged by the color of their skin but by the content of their character." With this statement Martin Luther King, Jr. opened the season of civil rights claims by African Americans at the end of the 1960s in the United States. Claiming the right to equality regardless of the color of the skin was in those years a kind of mirage, an illusion, an utopia:

a dream, precisely. Unexpectedly, Martin Luther King's dream came true, and without his having started from a position of power. "I have a dream" became a slogan, the symbol of how seemingly impossible goals can be reached. If Martin Luther King chose to incite his followers with the sentence, "I have a dream" and not, "I have a five-year plan," there must have been a reason: men need to share a dream to let their light shine.

In history more than once simple men were seen to make great projects come true with the mere strength of their aspiration. Gary Hamel, management professor, in his book, *Leaders of the Revolution* [4] gives several examples: "How many times has revolution been made by kings? Nelson Mandela, Vaclav Havel, Thomas Paine, Mahatma Gandhi: Had these men any political power? They didn't but they subverted the course of history thanks to passion."

Anatole France, Nobel prize winner for literature in 1921, singles out the stubborn prosecution of a dream as the way of obtaining great results: "To make great strides, we must not only act, but also dream, not only plan, but also believe" [51]. Peter Senge [5], an expert in social systems, believes the role of single people and minorities in determining change to be historically fundamental: "I don't know of many examples in history where significant changes were led by majorities and I see no reason to believe that this will be any different."

The American writer Carl Sandburg believed that nothing can happen without having been dreamed first: "Nothing can happen unless first a dream." The great cartoonist Walt Disney recognizes in dreams the capacity of liberating energy: "If you can dream it, you can do it." The German writer Hermann Hesse describes dreams as the source of the strength needed to overcome obstacles: "It's always difficult to be born [...] the bird struggles out of the egg [...] one's dream has to be found, to make the road easier." Jim Morrison, one of the greatest rock singers of the sixties, the leader of The Doors, maintained that dreams can liberate unexpressed potentialities: "Each of us has a pair of wings, but only those who dream learn to fly" [55].

When the focus of our attention moves from people to companies, the dream retains its evocative force, its propellant drive, and its ability of liberating energy. Kawamoto [6], the president of Honda Motor Corporation, in explaining the founder Soichiro Honda's "five commandments," underlines the central role of dreams in a large company's management.

We want to go on being a dream animated company, a company always young in spirit. Dreams—or ambitions—are the driving and positive force motivating us. In our existence we are urged to meet new challenges and never fear failure. In order for our dreams to become reality, we persist until all obstacles are overcome. In this research, we challenge ourselves as well as those around us. When our dreams finally become reality, we'll feel truly satisfied. [6]

The first who believed in the power of dreams was first of all the founder of Honda; Soichiro Honda didn't call his first motorcycle, his first creation, "Dream," by accident.

Many companies, in various fields, availed themselves of the image of dream in mass communication. We just quote a few. "Don't stop dreaming" (Sky). "The power of dreams" (Honda). "Dream Ideas" (Panasonic). "Long live dreams" (American Express, 2002). "If we didn't have dreams, we couldn't make them true" (BMW, 2007). "Nothing stops who decided to dream out of any constraint" (Audi, 2004). "We hold your dreams together" (Trenitalia, 2003). "I feel the hero of a dream" (Vodafone, 2002). "New Class S: The dream goes on" (Mercedes Benz, 2005). "To all who have a dream, we dedicated a custom made bank to make it true" (Unicredit, 2003). "The largest assortment of dreams" (Flou, 2006).

Mercedes Benz takes the idea of dream still further: "Dreaming. I see space and infinite dimensions through which my mind navigates. When I am awake I dream, I think ahead, I look ahead. I close my eyes and I see images and ideas flowing, coming to meet me. Darkness helps, it doesn't frighten me" [7]. In conventions and meetings the topic of dreams is often used in names: "Sign, dream, design" was the slogan the BMW group used in December 2001 for the presentation of the new Mini Cooper. A film director such as Steven Spielberg, an expert on imagination, in naming his motion picture company made dream his trademark: "Dreamworks."

The power of dreams comes from their peculiar ability to liberate energy. D'Egidio [8] singles out two sources of "energy" a company may avail itself of: the past, on one hand, the historical energy, constituted of accumulated experience, acquired knowledge, company climate, values, and rules; and the future, for the energy to come. When a company is born, the latter is necessarily the ruling one: "It is made of all the dreams, the imagination, the goals, the thoughts and the expectations. It's the energy created by a deep desire of realizing something new, different, exceptional, unique and irresistible."

Why do major international companies use dream as a motivational factor and evoke it in their advertising campaigns? The answer is that dream plays a fundamental role in unleashing energies that wouldn't otherwise develop and triggers positive emotions in imagining wished-for scenarios coming true. Dream can become the guideline, the line of march in someone's life, ambitions, and efforts, even when the dream appears unrealizable. On the contrary, pursuing one's dreams seems one of the greatest gifts nature put aside for mankind. "The greatest single human gift is the ability to chase down our dreams." This is what professor Allen Hobby states in the Steven Spielberg movie, *Artificial Intelligence* (2001).

The well known sociologist Alberoni links dream ideals, personality, and feelings.

> Extremely different skills are required to become a great entrepreneur, a great scientist, a great artist and a great political. And they appear extremely early. At the age of seven, it's already plain who can draw beautifully, who can do business, who can compose music, who is endowed with mathematical genius. But all these people, so different one from the other, have something in common. Each of them, in his domain, will achieve great things only by expressing himself fully, with his dreams, his personality, his life. His special abilities—musical, entrepreneurial, political creativity—are just the instruments for expression. An entrepreneur can create a vital and successful enterprise only by embodying his deepest feelings, his ideals, his spirit's richness in it. This enterprise is his novel, his symphony, his Sistine Chapel. It's the materialization of his mind. [9, p. 77]

Sergio Bambarén, an Australian writer, states: "We are as great as the dreams we strive to make true, and no matter what we meet along this journey called life. If we pursue our dreams with all our heart we'll understand the true meaning of our existence and we'll be almost sure to reach the goal we set [10]. "And after making one dream true, we pursue another one, as Hermann Hesse reminds us: "There's no everlasting dream. Each dream gives way to a new one and we mustn't try and keep any back." "When I dream, I live" (Native American chief): this is what appears in block capitals on the wall of the main stairs to Satel Group's offices in Pordenone. The proprietor, Francesco Regeni, when asked the reason for that choice, answered that this sentence echoes the sensibility of the main part of the company workers.

Carlo Talamo, who died in 2002, was an importer of Harley-Davidson and Triumph motorcycles. He maintained that an entrepreneur, like a child, should not think, but do things, pursuing his dreams: "Building an enterprise means strongly believing in it. Difficult moments are an integral part of a company's development. [...] As Morandi (an Italian singer) says "One of a thousand makes it, but how hard the climb is. But I am an eight year old child, and just as I did then, I dream of being a fireman, of driving a race car, of having a big chrome motorcycle making a lot of noise. Children don't think; they just do things. So, if we have a dream, we have to pursue it" [11, pp. 148–150].

Mark Fisher, who sold millions of copies with his bestseller *The Millionaire* [12], described how wishes and dreams shape people's existence, because man unavoidably becomes what he thinks every day. "At the base of each fortune is the faith of a man who believed in himself, who believed in an idea, in a dream, as crazy as it may have appeared to others." Victor Hugo would say: "No army in the world can stop an idea when its moment comes" [52]. Oscar Wilde comes to the point of saying that "An idea that is not dangerous, is unworthy of being called an idea at all" [53]. Oliver Wendell Holmes Jr. knew well the capacity an idea has of provoking irreversible changes, "Man's mind, once stretched by a new idea, never regains its original dimensions" [54].

Why are dreams important? Because, "But only in their dreams can man be truly free. 'It was always thus and always thus will be,'" says the famous actor Robin Williams as John Keating in the movie *Dead Poets Society*.

Myth as Means of Transformation

Group orientation and motivation are fundamental in change management. But in order to mobilize groups, powerful aggregation myths are required, endowing the group's actions and operations with meaning. Therefore, the essential passage for the management of transformation processes is the formulation and sharing of myths.

The word "myth" is derived from the Greek word *muthos* essentially meaning a string of words bearing sense. It's a speech, a public speech. It means the contents of these words, too, a thought. At the time of Homer's epos, it acquired the quality of fiction, and became a made-up story, an imaginary story, a tale, or an allegory. Myth opposes reality, intended as an obstacle to imagination, but is a true story at the same time. At the end of the nineteenth century, myth was an idealized representation of a past condition of humanity and of its origin. In the twentieth century a decisive role in representing a collectivity [13] or an individual [14] was ascribed to myth.

The process of expression of an individual as a unique being is a personal replica of humanity's great collective journey emerging from the primeval undifferentiation. This journey is represented, in primitive collective imaginary systems, by humanity's foundation myths [15]. In these great mythological–religious sagas humanity's process of development is described, with strikingly analogous topic and structure characteristics. In all foundation myths (the Judeo–Christian one, the Babylonian, Roman, Greek, and Sumerian ones, and so on) the same shared foundation phases are described: at the beginning the world appears in an undifferentiated and chaotic state. In a second phase, a function of separation and order in the primeval chaos appears, personified by an rebelling hero: Prometheus, Marduk, Gilgamesh, and so on. Then comes a regressive phase, in which the hero is temporarily defeated (Prometheus chained, Jesus going down to the underworld, and so on) and finally the process sets on its journey again and the hero prevails for good.

These phases correspond exactly, as Neumann [16] proved, to an individual's development, in his exiting the undifferentiated state of fusion with the mother, to his heroic facing the problems issued from separation and his giving in to the ensuing depressive and regressive spurs, finally arriving at his emerging as a differentiated and self-sufficient individual. The foundation myths are an echo, a sounding board, on the collective imaginary level, of this process, typical of humankind, archetypically predetermined and forever unfolding in the same topic mode.

Myths carry out the typical function of ancient religions: binding together (*religare*, in Latin,) the elements of reality in a comprehensible whole, explaining and arranging nature's terrifying and uncontrollable reality. Myths are appropriate systems for the explanation, communication, and signifying of shared experience, allowing man to face the first hard phases of contact with the chaotic elements in the environment [15].

Myths allow the discovery of a new way of understanding the whole and the relationship of the self inside that whole. Myths are the main vector of our migration and transition experience as well. Social transition needs myth as a means of transformation [17]. Myths determine the interpretation of the evolution in environment and bring it to transformation.

In today's life the existence of an organization without a shared myth is totally unimaginable. These myths originate in an instinctual fear of change. Their expression assumes several shapes, but all of them essentially illustrate the regulation of the relationship between the organization and the institution, explain the problems involved with change and "migration," and indicate the intensity and importance of the fears implied in evolutionary processes [17].

Many examples of social transition exist, especially in the kingdom of myths and stories. The expulsion from the Garden of Eden, the construction of the Tower of Babel, the journey to the Promised Land, to mention just a few, can be quoted. Obviously transition is always linked to the fear of the ensuing collapse of the actual order, the new shape of which is not defined yet. The expulsion from Eden speaks about life, about transition and the need for transformation [17].

Another meaningful concept is the "myth container" vehicle, allowing transition and possibly helping transformation as well [18]. Many historical examples can be quoted. The funerary ceremonial boats used to carry Pharaohs are an example. This kind of boat was used to carry to the necropolis a sort of catafalque, a platform adorned with palls supporting the coffin with the dead man's mummy. Terrestrial and fluvial boats were used. The fluvial boats were needed to cross the river separating the city of the living from the city of the dead. The journey would start from a purification basin where special funerary rites were performed on the mummy. In most cases, the transportation was done on the sand and in this case the boat was purely symbolic and pulled by a draught of oxen, or supplied with wheels [17].

In the Bible, the first and most ancient prototype of container or holy construction was the Ark, thanks to which Noah, his family, and the animals survived the cataclysm of the Flood. A further example of the "myth container" vector, allowing transformation, is the Ark of the Covenant, the symbol and the cornerstone prophecy in Jewish tradition. The Ark of the Covenant was the container where Israel had placed the Torah Tablets, after receiving them on Mount Sinai. The Ten Commandments were engraved on them. The Ark was carried for all of the 40-year journey through the desert, and came along with Israel in the long years of conquest of the Promised Land, until it was finally placed in the temple King Solomon built. The Ark was made by a case two and a half cubits long (one cubit = half meter) and one and a half cubits wide and high. When Israel set up the camp, a special tent made for the purpose was set up in the center and the Ark placed inside. It consisted of two main pieces: a parallelepiped underneath and a cover to close it, symbolizing the earth and the sky. Even if in nature earth is spherical (as all celestial bodies) and its movement elliptical (the circle being a particular case of an ellipse), according to cabalistic tradition the shape spiritually more suitable to represent the earth is the cube. In other words, today's universe is said to be ruled by spherical shapes, and the future one (the "new skies and the new earth") will mostly be inhabited by cubic shapes. This transformation holds the secret of the passage from a circular time (repeating itself, as in the Myth of the Eternal Return) to a rectilinear time, leading man toward a goal in all respects dissimilar from the point of departure. The conception of history as a series of events ferrying men from a less perfect state to a better and better one is one of the innovations of Jewish thought, and it has become an integral part of Western thought. Today it permeates both the laic and worldly concept of "progress," and the subtler and more refined idea of "evolution." From a symbolic point of view its representation is the transformation of the spherical shape of physical space into a cube.

Myth sets the borders of the inclusion–exclusion from the group and supports the feeling of belonging to a whole. Myths therefore acquire the value of refoundation of the origin, of the world's order and purpose [19]. If dream is the imagination of intimacy, myth is the imagination of what is social, public, and collective. Dealing with collective imagery, René Kaës [19] underlines two functions:

- An exploratory imagination, related to the primary processes of the portrayal of the unknown: this is a group's dream.
- An explicatory imagination, aimed at creating a shared and agreed-on representation of the group members' ego: this is the myth.

The group's dream and myth are therefore two forms of collective imagination: the first is the imagination exploring the unknown, and the second is the imagination explaining it, the explicatory imagination.

Finally, myth is the explicatory imagination of the members of a group, an imagery allowing someone to make an initiation journey, on a path out of time, in order to experience the hero's journey and the birth of self, a journey tuned to the rhythm of the great mythological sagas brought back to life in the dreams of individuals themselves [15].

Vision as the Imagination of the Organization

So, myth is the collective imagination explaining, communicating, and giving meaning to individuals' shared experience. The more society lives through transition, the more important myths are, and truly become a vehicle of transformation. Myths emerge from the instinctive fear of change, and decide the interpretation of the evolution in the environment, thus leading to transformation.

When the focus of our attention shifts from society to organizations, and especially to companies, the function of myth—as society's imagination—is carried out by vision, which can be considered to be a company's imagination. The very definition of vision allows us to consider it as something pertaining to imagination. In fact in the Italian *Zingarelli Dictionary* vision means: "a perceptive process through which the knowledge of the outer world is achieved [...] the act of seeing [...] the perception through sight of super-natural realities [...] description of things that have been seen, both in dream or sensibly [...] visual perception of images or events which, although not real in themselves, are originated in reality, pertain to reality and may become real." Inside an organizational context, the company's vision acquires the meaning of shared dream. The relationships among dream, vision, and myth are given in Table 4.1. Three different classes of imagination are offered:

- Dream as the heart of hearts' imagination
- Vision as the organization's imagination, or shared dream on an organizational level
- Myth as society's imagination, or shared dream on a social level

Dream is an exploratory imagination, that is, an enquirer of the unknown, whereas vision and myth are "explicatory" imaginations, and explain hidden elements.

TABLE 4.1

Classes of Imagination[a]

Domain of Imagination		
Individual	Organization	Society
Dream	Vision	Myth
Heart of hearts' imagination	Organization's imagination	Society's imagination
Exploratory imagination	Explicatory imagination	Explicatory imagination
Individual's dream	Shared dream at organiz. level	Shared dream at social level

[a] From De Toni, A.F. and Barbaro, A. (2009). Imprenditorialità e sogno imprenditoriale. In L. Cassia, M. Kalchschmidt, and S. Paleari (Eds.), *L'imprenditorialità: Pensiero, elementi, contesto*. Bergamo, Italy: Bergamo University Press, Sestante Edizioni. [20]

Gary Hamel [4] agrees with the companies' need of having and pursuing dreams: "In the progress era, dreams weren't much more than reveries. Today, as never before, dreams are the anteroom of new realities. Our collective selves—our organizations—must learn to dream." The idea of a shared dream brings about an apparent contradiction between individual and collectivity. There is the need for an individual project, a unique, personal, unrepeatable project, but on the other hand there is the need to belong to something. So, should an organization bet on individuals or on groups? The two Swedish professors Ridderstrale and Nordstroem [21] suggest "both."

> Not only are we individualistic creatures, we want to belong as well, we need sense and goals as well. This is why the company of the future won't be either individualistic or collectivist, but both: strong personalities keep together if a dream is shared between them. But mind, the important thing is that it should precisely be a dream. The subordinates in these companies are precisely living a religious experience. They know they are part of a group with a project, but at the same time contribute to that vision themselves. The beacon, in both cases, is dream.

The leader's dream, an individual's imagination, must be acquired by the company's group and become the vision of the enterprise, the organization's collective imagination. The power of the vision on the company level equals the power of myth on the social level. As myth is a means of transformation from one social situation to another one, vision leads the company group to the attainment of the goal, marking the line of march, fixing the borders of inclusion in the group, and supporting the feeling of belonging to the company. Vision, like myth, has a grounding value for the origin and goals of the group, and becomes the ideal place where the individual's journey of development acquires meaning.

D'Egidio [8] defines vision as "a powerful and irresistible image of what we want to create in the future. Aristotle stated that man thinks and acts on the base of images [...] Vision is much more than an idea: it's a force in people's heart, a force endowed with incredible power." According to the author: "The creation of a shared vision brings people to be aware of their dreams and to share other people's dreams. A strong feeling of sharing and cohesion ensues."

This "power" of vision is recognized by Brunetti and Camuffo [22] as well, according to whom a company must be managed and governed by people with a vision, a line of march. Embracing the thought of the psychologist Quaglino, these two authors state that men with a dream and the ability of transforming it into a need to participate in the construction of something special should lead companies.

Sharing a meaning is considered fundamental by Warren Bennis [23] too:

> People would like to spend their life in a cause they believe in, instead of dragging themselves along a second-rate existence devoid of meaningful ideals. [...] It's necessary to create a shared aim, because people really need an aim; a meaningful aim. This is the reason why we live; and I think that a company's strength will stand in that very shared aim. With such an aim, a stimulating and shared aim, everything can be reached.

The sociologist Alberoni writes, as well: "We expect a true commander to endow our actions with meaning" [24, p. 21]. A shared aspiration ensures the energy needed for success; this is what Prahalad [25] maintains:

> Companies need a widely shared aspiration [...] This is the fuel that drives the engine [...] Strategic architecture provides a company with the direction, but it needs to have the emotional and intellectual energy to make the journey. It needs shared aspiration, which allows the company to stretch itself beyond its current resources; one that provides a sense of direction, a sense of common purpose, a single-minded and inspiring challenge which commands the respect and the allegiance of every person in the organization.

To sum up: in change processes, the *leitmotif* of project management, a strong vision is needed, a vision capable of directing the group, activating its motivation, and liberating its creative energies.

SHARED VISION AS THE FUNDAMENTAL LEVER FOR CHANGE

We have seen how the dream is the driving force of an individual's creative push. Sharing dreams, intended as an individual's imagination, in a group is the fundamental condition in order to make it come true. The dream must become a shared dream, the organization's vision. The power of the vision, intended as the imagination of the group, inside the group, has the same power as myth on a social level. As myth is the means of transformation from a social situation to a different one, shared vision directs the group during change from one condition to another.

On the simplest level, a shared vision is the answer to the question: "What do we intend to create?" Exactly as personal visions are representations or images people carry in their heads and hearts, so shared visions are representations adopted by the people in an organization. If well developed, shared vision is a very powerful force.

According to Peter Senge [26], "A shared vision is not an idea [...]. It is rather some strength in people's heart, a strength with an impressive might. It may be inspired by an idea, but once it goes beyond it—if it is engaging enough as to acquire more than one person's support—it is not an abstraction anymore. It is palpable. People start seeing it as if it existed. In human affairs few things, if any, are as powerful as a shared vision." The author explains how energy must be sought inside, maintaining that the persuasion of being able to realize something great is the source of extraordinary strength:

> We must stop trying to understand what we have to do by looking at what we did in the past [...] we must start really looking into our heart, to find what appears truly possible to us. That literally means enacting a process of change inspired by a vision. The source of energy is in our inner persuasion that some thing can be done. And we might be talking about a product never produced before on a large scale: all historical data say no, but our heart says yes.

A group without a vision is like a ship without a destination. A worker without a vision is not aware of the final goal of his work. Three stonecutters, working on the preparation of stones to build a castle, were asked, "For what reason do you work in this place?" The first answered, "I'm working because this is how I'll be able to eat." The second replied, "I'm working to get the stones ready for the construction of this wall." The third stonecutter, jumping to his feet and swelling out his chest, gave this answer, "I am working in order to finish that huge castle that will be built over there" [27, p. 37]. Sharing the vision is the condition needed to participate in the joy of producing, of creating.

Eric Motley (who was in charge of the selection of former U.S. President George W. Bush's advisory board members at the White House, a key role in the president's entourage) explained, in a 2002 interview, that the key words of success are vision, group, and shared values. "The first step is to shift from the idea of 'I' to the concept of 'Us' and to understand that one is too small a number to do great things. [...] The desirable qualities to be able to give a constructive contribution to a team are: strong character, loyalty to the team, broadmindedness, wide-ranging interest, honesty and self-confidence." An idea the human resources expert insists on is the need on the group's part of sharing common values and having a vision. "It's important to understand where the team is going, make sure to have the right people for the journey and subordinate one's goals to the team's."

Valdani [28] says: "The formulation of a vision doesn't warrant a profitable adventure in the future world, but without vision no journey can start." And in order to warrant a "profitable adventure," as Valdani calls it, the organization must share the vision.

THE CONSTRUCTION OF THE SHARED VISION

Man feels the need for sharing: only by sharing a group's values and culture may a person be accepted. Sharing therefore becomes first of all a mechanism of acceptance in social systems, as writes Gharajedaghi [29, pp. 84–85]:

> The shared image is the main tie between the members of a human community, and it encourages the rising of the required conditions for any significant communication. The degree with which the single person's image coincides with the community's shared image, determines the former's degree of participation to the community itself. It's the shared image we refer to when we speak of a people's culture. This embodies a people's experience, beliefs, attitudes, ideals, it's the final product of his history and the manifestation of his identity; man makes his culture and his culture makes the man.

If we cut any picture, each part shows only a fraction of the whole image. If we divide a hologram, each part represents the whole image, intact. Likewise, if the hologram goes on being divided, no matter how small the parts, each piece will still represent the whole image. Similarly, when a group of people shares the vision of an organization, each of them gives a representation of the organization. Each shares the whole of the responsibility, not only in what concerns his own part. The "pieces" composing the hologram are not identical, but each of them represents the whole image.

Senge [26] states: "When the pieces of a hologram are summed up together, the image of the whole doesn't radically change. After all, it was in each piece. But nevertheless the image becomes more intense, more vital [...]. As the shared vision develops, it becomes at the same time 'my vision' and 'our vision'."

Hans Juergen Warnecke and Manfred Hueser defined the ideal organization as a fractal. Starting from this definition, Savage [30] comes to the conclusion that in such a structure every component is as independent as needed to take one's decisions autonomously, but at the same time the decision meets criteria the whole organization shares. Or rather, referring to the hologrammatic model, the part is in the whole, the whole is in the part.

Adriaplast, a chemical industry part of the multinational Solvay Group, indicates sharing as a must for vision: "The vision in Adriaplast is clear, obvious and measurable; challenging; strong and accessible to everyone; shared; spread to all levels" [31]. Bessone concludes saying: "Our Vision is an image of the future we want to create, described at the present tense, as if already happening."

In order for a vision truly to become a powerful strength, it needs (see Figure 4.1):

Words, for it to be exciting
Actions, for it to be real, tangible
Relations, for it to be shared with everyone

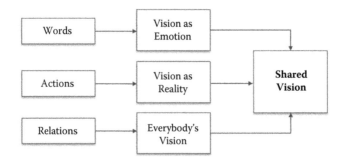

FIGURE 4.1
Construction of the shared vision [32, p. 386].

Vision as Emotion

Vision must first of all be an emotion. Without this component motivation is reduced to a short-lived activity. A leadership lacking the required contact with individuals ends up withdrawing from real-life situations, unable to have a grip on its people's lives and aspirations. "An act of persuasion is needed; a firm and enthusiast engagement must be created for the organization and for the people working in it. Focusing his collaborators' attention on the vision, the leader is acting on the company's emotional and spiritual resources, on their values, on their engagement and their aspirations" [33].

According to Corrigan [34] in Shakespeare's works leaders succeed in excitingly conveying their vision as emotion. Here is a quote, as an example of vision as emotion in Shakespeare's theatre. The example concerns Henry V, addressing his men before the battle of Agincourt, where the English army must face an army, the French one, ten times as large [50].

> We few, we happy few, we band of brothers
> For he today that sheds his blood with me
> Shall be my brother; be he ne'er so vile,
> This day shall gentle his condition;
> And gentlemen in England now abed
> Shall think themselves accursed they were not here
> And hold their manhoods cheap whiles any speak
> That fought with us upon Saint Crispin's day.

Henry V ends his transfer of emotional energy to his men with the following statement, a very effective one, on our opinion: "All things are ready, if our minds be so." According to Corrigan [34], "Shakespeare's Henry is endowed with a very powerful vision, capable of producing deep transformation: after hearing the king's speech, the men he is addressing become different soldiers, become part of a larger entity. Henry shows managers that if they want their people to be a cut above, new stimuli and a new enthusiasm are required."

The first characteristic management must try to achieve is therefore the ability of formulating a clear vision. In the 1960s, John F. Kennedy brought the whole nation together with the renovation of the American myth of the new frontier, and moved it from the West to Space: the challenge consisted of sending man to the Moon within the end of the decade. People had to be emotionally involved.

One must somehow be an artist (Hammer and Champy [35, p. 173]): "Creating the vision of the organization [...] requires a kind of artistic ability, because a vision is an image without details." According to Schein [36], the evocation of stories, myths, and legends is as important as formal vision. And Kets de Vries and Miller [37] note, "This homogeneity of points of view may be strengthened by myths, legends and stories about the organization, allowing members to identify shared symbols, to reach a sense of community and to give birth to shared fantasies. These fantasies could be about the origin of the company, the story of its evolution, the difficulties overcome, the rites of transition, and involve all inside and outside relations in the organization. According to Mintroff and Kilmann 'corporate mythology is the spirit of the organization and is spread on all management levels.'"

Vision as Reality

Vision must not only consist of emotion; it has to include reality as well. Obviously, if vision, as energizing and emotionally involving as it may be, is committed to paper only, it cannot be shared. Actions are needed on management's part to show how vision is something real, to be used daily. Bennis [23] writes:

> To convey a vision something more than words is needed. It's not a question of bright speeches, of commanding notes or glittering plates hanging on the wall. It's a question of living through the vision, day by day, of interiorizing it and granting autonomy to other members of the organization, in order to allow him to implement and make the vision true in every single action [...] If there ever was some truth, it is that actions are more important than words.

According to Lowney [38, p. 89]: "Leaders can influence other people with their example, their ideals and their teachings." Peters [39, p. 401] reinforces the same concept: "Posters expressing the vision and company values charts can really be useful, but they can bring about the opposite effect. In fact, the risk is to hamper and mimic the motivational process if the vision and the values are just proclaimed and are not lived though in a convincing manner." In short, leaders should set the example and show the steps to be taken for an individual to contribute operatively to that vision. This is the only way a vision may be perceived as real and actually become so.

Everyone's Vision

Finally vision must be born, develop, grow, and change through the relations inside the company, and become therefore everyone's vision.

Bernardi and Muffato [40, p. 198] state: "The vision statement cannot be the result of a top-down process or be an 'official' vision reflecting only few people's opinion. Furthermore, the attention it's given cannot last just a few moment[s], when the vision is formulated, and then be forgotten. On the contrary, the vision statement must be confirmed by facts all the time."

With special attention to the complex adaptive systems, as a company, Olson and Eoyang [41, p. 73] write that the vision emerges out of the inter-action between the agents of the system:

> Vision allows the members of a complex adaptive system to know who they are, what they can do well, and what is the direction they want to move to. For the development of a vision in an adaptive complex system, it's essential to understand the actual dynamics and to allow system members the con-struction of future possibilities. The vision emerges from the place where order and disorder meet, in the rich interaction of experiences, thoughts and connections of the system's agents.

A vision for everyone must take into account the different demands that are present in the company (Ciappei and Poggi [42, p. 163]): "From this viewpoint the organizational culture must be conceived as the result of an aggregation process of the different cultural demands present in the company, since this is the only way to involve all who can generate the shared vision." In this respect, differences, countercultures, and the pres-ence of protesters is very important in order to avoid the vision being flat-tened under the prevailing position.

On final analysis, a manager's duty is a complex one, full of contradic-tions, always on the edge between the subordinates' emotional involvement and contact with reality, not to be lost anyway. Peters [39] writes: "The most effective political or corporate leaders urge others to act—and develop—on support of a cause considered worthy by both the leaders and the subordi-nates. A leader's duty is to enrich empowering vision and at the same time to keep in touch with his staff in order to make sure to be on the same wavelength required in the real world where the vision is implemented."

NEW MANAGERIAL MODELS IN ORDER TO FACE COMPLEXITY

Shared vision is therefore the engine for change on the organizational level. It implies first of all the activation of self-organizational models.

TABLE 4.2

Comparison between Classic Model and Complex Model[a]

	Managerial Model	
Characteristics	**Classic Model**	**Complex Model**
Organization	Traditional (hierarchical)	Self-organization (nonhierarchical)
Environment	Mostly steady	Turbulent
Future	Partially predictable	Unpredictable
Success	Balance and stability	Unbalance and change
Decision-making processes	Determined	Undetermined
Tools	Logical and analytic	Intuitive and analogy-based
Learning	Historical data extrapolation	By attempts
Management	Rules	Shared vision
Impact of inside differences	Negative	Positive
Managers' prevailing role	Planning and control	Creation of learning and innovating context
Managers' prevailing trend	Executive	Exploratory
New strategies formulation	Top down, backing agreement, order, harmony	Bottom up, backing conflicts, disorder, discordance
Goal	Stability (complexity reduction)	Flexibility (complexity absorption)

[a] From De Toni, A.F., Comello, L., and Ioan, L. (2011). *Auto-organizzazioni. Il mistero dell'emergenza dal basso nei sistemi fisici, biologici e sociali.* Venice: Marsilio. [43]

Self-organizational models are nonhierarchic organizational models. Among those proposed in specific literature we may quote: the circular model, hologrammatic model, cellular model, and holonic model [43]. In this very book these models are described in Chapter 5, Luca Comello's contribution. Traditional hierarchic organizational models are overcome within the framework of the transition from a traditional managerial model to a so-called complex managerial model (see Table 4.2).

Roberto Costantini, in the Italian epilogue to Pascale's book [44, p. 388], highlights the main differences between the classic model, suitable for simple situations, and the complex model, applicable in complex situations:

> The old paradigm inclines to the construction of stability, predictability, and little risk (fail-safe world), while the new paradigm is based on the assumption that the future is unpredictable and turbulent and that it is therefore important for instability to be managed an all options to be

kept open (safe to fail world). Essentially, the old paradigm is handier to manage, more reassuring, on the short term, for investors, but certainly less adequate to the reality of the world around us and to long term development.

Adopting the complex model implies the idea of abandoning classic reductionist ideas. Avoiding reductionism means avoiding the attitude that, as a rule, often shaped companies' behavior: the attempt to aim at a simple representation of an otherwise complex reality in order to encourage decision making. The classic model considers an organization as a simple structure, set in a stable environment and in a predictable future. Success under this condition ensues from balance and stability. The complex model, on the other hand, is based on the idea that an organization is an adaptive complex system, set in a turbulent environment and in an unpredictable future. In such a situation, success comes for imbalance and change, as does survival for adaptive complex systems: according to Pascale and others [45]: "balance is dead."

Decision-making processes carried out according to the classic model are determined, meaning that they follow a precise procedure, whereas in the complex model they are undetermined, continuously discussed, and subject to modification [...] Gozzi states (Ciappei and Poggi [42, p. 163]): "The new decision-making process) is a research journey, tortuous, nonlinear, full of unpredictable and unknown events, ambiguous and contradictory here and there [...] generation and manipulation of knowledge rather than well-established procedure, under the constant control of the decision-makers, and the guide of calculation and organization."

According to Pascale [45]: "Management per goals is not very useful. We don't want static targets, but a process aimed at steadily making things better." The classic model bases management on rational rules, considers the presence of internal dissimilarities to be negative and assigns to management a prevailing task of planning and control. In order to give the best guarantee for managing co-ordination, the complex model leverages the shared vision obtained through activation of social processes; it furthermore considers positively the presence of inside differences as a source of innovation, and assigns to management the task of creating favorable conditions for learning and constant innovation.

The new approach to management is much less comforting than the classic one, doesn't reduce anxiety, but is much more dynamic and useful in turbulent times. Top managers in large organizations are faced with the following alternative.

- Allowing the emergence of new strategies. Conflicts, disorder, and disharmony will follow this process. A new strategy might emerge, or maybe not.
- Insisting on adhesion to the formulated plan. By the application of direct forms of control to situations with an unpredictable ending the risk of conflicts, disorder, and disharmony is considerably reduced. As a result the emergence of a new potential direction will be blocked. What appears as the safest alternative, the second one, actually is the most dangerous one, not taking into account the dynamics of the game.

As for the final goal of the two models, Kenwyn Smith of Wharton School maintains that whereas the classic model is stability-oriented, the complex model is flexibility-oriented (Pascale [45]). The difference between stability and flexibility emphasizes two radically opposite views of organization management. Orientation to stability highlights balance, therefore assigning all resources to the maintenance of a predictable situation (a world safe from mistakes) becomes especially important in such a scenario. According to the second orientation, energies must concentrate on flexibility, highlighting the importance of keeping one's options open (a world of "safe" mistakes). In such a scenario, the basic hypothesis is that the future is unpredictable, rather than predictable.

This idea resembles the concept of stability and resilience studied by the Ecologic School [46]. In today's ecologic theory, as proposed by Holling, May, Ewing, and others, resilience can oppose simple stability. Stability is the ability of an ecosystem to return to a stable situation after a temporary perturbation: the quicker it goes back and the smaller the fluctuation in comparison to the rule of its steady status, the more stable the system is. Nevertheless another property, called resilience, turns out to be more important whenever the point of view of the maintenance of ecosystems over the long term is adopted: resilience is the measure of the ability of the system to absorb change and perturbations, and to find stable solutions of state with respect to a series of fluctuations covering an ample range of directions as well.

Stability deals with reduction of complexity and flexibility with the absorption of complexity. The traditional model is not wrong, but insufficient, as Savage [30] states: "When the future is like the past, organizing through routine makes sense. But when round spheres, unexpected harmonies and kaleidoscope technological changes are everyday occurrences, the implementation of a strategy aimed at complexity and variety is required."

THE PROJECT MANAGER'S NEW ROLE

In the complex management model, the project manager's leadership may seem less important, but it is not so. In a logic of self-organization, according to Vicari [47, p. 147]: "The leader's functions increase, instead of decreasing. In fact self-organization, as physics teach us, can also happen spontaneously, but always under certain conditions. The manager's task is, therefore, the creation of such conditions […]."

Anderson [48, pp. 216–232] maintains management should supply the outer energy required for the self-organization of complex adaptive systems:

> Self-organization will not apply without a continuous energy flow inside the system. Nevertheless the studies about how managers bring energy to the organization have been separated from the studies as to how structures emerge and evolve. The effort level in organizations changes if managers push them toward new activities, bring new challenges and goals to the members' attention, form and break connections inside and outside, change the awarding systems. […] Understanding causes and conditions of an input of energy in the system in a network under agents' evolution is an important topic for future research.

And Gharajedaghi [29, pp. 84–85] adds: "Power is like knowledge. It can be duplicated. The conceptualization of power as an entity summing up to non-zero is the critical step in order to understand the essence of empowerment and the 'many-minded' systems management. Empowerment is neither, therefore, power renunciation nor power sharing. It's power duplication."

Lowney [38, p. 89], in his book *Heroic Leadership*, underlines that "Practicing leadership means having an influence, having a precise vision, being persevering, infusing the required energy in others, being open to innovation and offering one's teaching." Lowney's thesis is that the real motivation is self-motivation: "Whether a person works in a large company or in total solitude, no mission will supply the right motivation unless it is one's own." Further on: "Most assuredly, a company's mission will enflame only the people who formulated it, and that because the very process of formulating it made it personally important for the people involved. Leaders and managers, therefore, should find the way to render this mission personal also for their subordinates. That is the key to create self-motivation." Lowney still quotes: "We learnt from experience that each man finds a greater pleasure and a greater stimulus in what he is able to discover in his inner self. It will therefore be enough to point, even with a finger, to the auriferous vein of the deposit and then let everyone start digging by himself" (Meissner [49]). The final conclusion the author gets to is that every leader should enable single members of the group to become his own leader.

In order for every member to become his own leader and help self-organization processes, the project manager must operate informally and act on intangible elements. Values (as ethics in relationship), culture (conceived as specific wealth of knowledge and notions organically bound in order to give a substantial contribution to a single person's or an organization's "personality" building), and language (meant as interaction codes, bearers of implicit meaning) need to be shared.

In short, the project manager's role changes from a reductionist role to a complex one, from "planning and control" to the "creation of context." A context where the true motivation is self-motivation, coming out of shared vision, obtained under the leader's example; the project manager supplies the energy for change. In order to manage the growing complexity, the goal should consist of participation and assumption of responsibility on everyone's part. Evenly shared, interconnected, self-motivated, and self-activated intelligence is required. There's no solution in the center. The future is on the outskirts.

CONCLUSIONS

If the driving forces of change for people and organizations are, respectively, dreams and visions, the project manager will on one hand encourage people's dreams in order to enhance their evocative drive, the propelling push, and the ability to liberate creative power, and he will also create a favorable context for the sharing of the vision in order to activate self-organization processes.

If management means "management of becoming," then real management is project management. Project managers enjoy a distinctive privilege: working on change. As Arthur Schopenhauer maintains, "Change only is eternal, everlasting, immortal."

REFERENCES

1. Lanternari, V. (1966). Il sogno e il suo valore culturale dalle società arcaiche alla industriale [Dream and its cultural value from archaic to industrial society]. In von Grunebaum and Caillois, *Il sogno e le civiltà umane [Dream and Human Civilizations]*. Bari, Italy: Laterza.
2. Lombardozzi, A. (2002). Le culture del sogno: Una prospettiva antropologica [The culture of the dream: An anthropologic view]. *Funzione Gamma Journal*, 10.
3. Fubini, F. (2002). Sogni in cerca di un sognatore [Dreams looking for a dreamer]. *Funzione Gamma Journal*, 10.
4. Hamel, G. (2000). *Leader della rivoluzione [Leaders in Revolution]*. Milan: Il Sole 24 Ore.
5. Senge, P. (1997). Through the eye of the needle. In Rowan Gibson, *Rethinking the Future: The New Paradigms of Business*. Milan: Il Sole 24 Ore - Libri.
6. Derisbourg, Y. (1993). *Mr. Honda as Told to Yves Derisbourg*. Milan: Lupetti.
7. Advertisement published in newspapers in November 2005.
8. D'Egidio, F. (1993). *Il sogno imprenditoriale. L'incredibile storia di un manager innovativo. [The entrepreneurial dream. The Unbelievable Story of an Innovative Manager]*. Milan: Franco Angeli.
9. Alberoni, F. (2002). *L'arte del commando [The Art of Command]*. Milan: Rizzoli, p. 77.
10. Bambarèn, S. (2000). *The White Sail*. Milan: Sperling & Kupfer.
11. Pacenti, G.C. (2000). *Imprenditori si nasce o si diventa [Is one born an entrepreneur or does one become so]*. Milan: Franco Angeli, pp. 148–150.
12. Fisher, M. (2001). *The Millionaire*. Milan: Bompiani.
13. Levi-Strauss, C. (1958). *Anthropologie structurale*. Paris: Plon. English edition: *Structural Anthropology*. Chicago: University of Chicago Press, 1983.
14. Lacan, J. (1986). *Il mito individuale del nevrotico*. Rome: Astrolabio.
15. Zanasi, M. (1996). *Archetipi e Caos in Chaos, Fractals Models*. F. Marsella Guindani and G. Salvatori (Eds.). Pavia, Italy: Italian University.
16. Neumann, E. (1978). *Storia delle origini della coscienza*. Rome: Ubaldini.
17. Gildenhuys, A. (2002). Creazione del mito, transizione sociale e trasformazione: Viaggio nell'età del sogno. *Funzione Gamma Journal*, 9.
18. Lawrence, G. (1991). *Won from the Void and Formless Infinite: Experiences of Social Dreaming*. Free Associations, 2, 22: 259–294.
19. Kaës, R. (2002). Sogno o mito? Due forme e due destini dell'immaginario. *Funzione Gamma Journal*, 9, June.

20. De Toni, A.F. and Barbaro, A. (2009). Imprenditorialità e sogno imprenditoriale. In L. Cassia, M. Kalchschmidt, and S. Paleari (Eds.), *L'imprenditorialità: Pensiero, elementi, contesto*. Bergamo, Italy: Bergamo University Press, Sestante Edizioni.

21. Ridderstrale, J. and Nordström, K. (2006). *Karaoke Capitalism*. Milan: Franco Angeli.

22. Brunetti, G. and Camuffo, A. (2000). *Del Vecchio e Luxottica*. Tutin: Isedi.

23. Bennis, W. (1997). Diventare leader di leader. In R. Gibson, *Ripensare il futuro. I nuovi paradigmi del business*. Milan: Il Sole 24 Ore-Libri.

24. Alberoni, F. (2002). *L'arte del comando [The art of the command]*. Milan: Biblioteca Universale Rizzoli, p. 21.

25. Prahalad, C.K. (1997). Strategie di crescita. In R. Gibson (Ed.) *Ripensare il futuro. I nuovi paradigmi del business*. Milan: Il Sole 24 Ore - Libri.

26. Senge, P. (1992). *La quinta disciplina. L'arte e la pratica dell'apprendimento organizzativo*. Milan: Sperling & Kupfer. *The Fifth Discipline*, New York: Doubleday, 1990.

27. Tateisi, K. (1992). *L'irresistibile spirito dell'imprenditore. La filosofia pratica di un manager*. Milan: Sperling & Kupfer, p. 37.

28. Valdani, E. (2000). *L'impresa proattiva. Coevolvere e competere nell'era dell'immaginazione*. New York: McGraw-Hill.

29. Gharajedaghi, J. (1999). *Systems Thinking: Managing Chaos and Complexity*. Boston: Butterworth-Heinemann, pp. 84–85.

30. Savage, C.M. (1990). *5th Generation Management. Co-Creating through Virtual Enterprising, Dynamic Teaming, and Knowledge Networking*. Boston: Butterworth-Heinemann, pp. 130–131.

31. Bessone, P. (2002). Vision 2000: *Le risorse umane nel contesto di una multinazionale. Guidare il cambiamento*, Palmanova, Udine, Italy, December.

32. De Toni, A.F. and Comello, L. (2005). *Prede o ragni. Uomini e organizzazioni nella ragnatela della complessità*. Turin: UTET University, p. 386.

33. Foster, R. and Kaplan, S. (2001). *Creative Destruction: Why Companies That Are Built to Last Underperform the Market—And How to Successfully Transform Them*. New York: Doubleday.

34. Corrigan, P. (2000). *Shakespeare on Management: Leadership Lessons for Today's Managers*. London: Kogan Page.

35. Hammer, M. and Champy, J. (1993). *Ripensare l'azienda. Un manifesto per la rivoluzione manageriale*. Milan: Sperling & Kupfer, 1994, p. 173. *[Reengineering the Corporation]*. New York: HarperCollins.

36. Schein, E. (1985). *Organizational Culture and Leadership*. San Francisco: Jossey-Bass, p. 82.

37. Kets De Vries, M.F. and Miller, D. (1984). *The Neurotic Organization*. San Francisco: Jossey-Bass.

38. Lowney, C. (2005). *Leader per vocazione. I principi della leadership secondo i gesuiti*. Milan: Il Sole 24 Ore-Libri, p. 89. *[Heroic Leadership*, Chicago: Loyola Press, 2003].

39. Peters, T. (1989). *Thriving on Chaos*. London: Harper Paperbacks, p. 401.

40. Bernardi, G. and Muffatto, M. (1992). La cultura organizzativa nei processi di cambiamento. In R. Filippini, G. Pagliarani, and G. Petroni (Eds.), *Progettare e gestire l'impresa innovativa. I nuovi percorsi per affrontare la complessità degli anni Novanta*. Milan: Etas, p. 198.

41. Olson, E.E. and Eoyang, G.H. (2001). *Facilitating Organization Change. Lessons from Complexity Science*. San Francisco: Jossey-Bass/Pfeiffer, p. 73.

42. Ciappei, C. and Poggi, A. (1997). *Apprendimento e agire strategico di impresa. Il governo delle dinamiche conoscitive nella complessità aziendale*. Padua: Cedam, p. 163.

43. De Toni, A.F., Comello, L., and Ioan, L. (2011). *Auto-organizzazioni. Il mistero dell'emergenza dal basso nei sistemi fisici, biologici e sociali.* Venice: Marsilio.
44. Pascale, R.T. (1992). *Il management di frontiera. Come le aziende più intelligenti usano conflitti e tensioni per diventare leader.* Milan: Sperling & Kupfer, p. 388 [*Managing on the Edge*, New York: Doubleday, 1990].
45. Pascale, R.T., Millemann, M., and Gioja, L. (2000). *Surfing the Edge of Chaos. The Laws of Nature and the New Laws of Business.* New York: Three Rivers Press.
46. Laszlo, E. (1985). L'evoluzione della complessità e l'ordine mondiale contemporaneo. In G. Bocchi and M. Ceruti, *La sfida della complessità.* Milan: Feltrinelli, p. 379.
47. Vicari, S. (1998). *La creatività dell'impresa. Tra caso e necessità.* Milan: Etas, p. 147.
48. Anderson, P.W. (1999). Complexity theory and organization science. *Organization Science: A Journal of the Institute of Management Sciences,* 10: 3, 216–232.
49. Meissner, W.W. (1992). *Ignatius of Loyola: The Psychology of a Saint.* New Haven, CT: Yale University Press.
50. Shakespeare, W. *Henry V*, Act IV; Scene III.
51. Anatole France. http://www.brainyquote.com/quotes/authors/a/anatole_france_3.html
52. Victor Hugo. http://en.wikiquote.org/wiki/Victor_Hugo
53. Oscar Wilde. (1891). *The Critic as Artist.* http://en.wikiquote.org/wiki/Censorship
54. Oliver Wendall Holmes. http://thinkexist.com/quotation/man-s_mind_once_stretched_by_a_new_idea-never/8182.html
55. http://www.bellefrasi.it/en/frasi-di-jim-morrison-sull-amore/

5

Self-Organized Project Management

Luca Comello

CONTENTS

PROJECT MANAGEMENT: ORIGINS AND DOMAIN

I work as a project manager in a company. I do my best to apply best practices in the discipline and adapt them to the context I work in, and respect the company's story, its culture, and its processes. But I'm interested in the theory of complexity applied to management too. So daily I'm faced with questions such as: is a project really just a sum of activities? Is it really possible to make a plan and then keep to the initial plan? And what about risk? Is time the linear sequence of activities governed by the deterministic relation "finish-to-start" I set in my software? Is it important to change the point of view about project management?

Complexity, in management generally speaking and in project management in particular, was and still is considered as a complication, thinking of the linearity of the folds in a sheet of paper, to be explained with an analytic approach, and not of the nonlinear weaving of people, relations, connections, and worlds, to be described with a systemic approach. According to this point of view, inside a company complexity represents the number of subsets that can be identified and may be considered according to three categories (a vertical one, the number of levels in the hierarchy; a horizontal one, the number of unities throughout the organization; and a spatial one, the number of geographic places). Outside, are the number of different elements the company has to deal with at the same time.

To deal with complexity as complication, the tools we know are essential. But when the unexpected comes true, when the treasure of possibilities amazes us once again, when all of a sudden discontinuity follows continuity, it's time to integrate them with new approaches. We are entering the era of emergence. In order to understand the meaning of this passage, we had better briefly observe what is happening in science.

THE ERA OF EMERGENCE

The birth of modern science can be traced back to the seventeenth century and to a fundamental choice: the basic assumption is that, inasmuch as studying nature as an organic whole is impossible, limiting the objects of study to linear and quantifiable phenomena artificially separated from the environment is necessary. In one word: simplification!

Thinking about it, this is our typical approach, both as project managers and in nonprofessional domains, when faced with a problem: we dismantle it, then reduce it to parts small enough to be solved easily, and sometimes we describe them with simple mathematical laws (we try to quantify the duration of an activity, the cost, the use of resources, and so on).

This methodological approach, called reductionism, is the starting point of some of the most unbelievable progress in the history of knowledge. The research nowadays covers a range of scales including over 60 orders of magnitude, from the unimaginably small dimensions of subnuclear physics, to the possibly boundless ones of cosmology. In little more than two centuries discoveries included the laws of gravitation regulating planetary motion; the laws of thermodynamics that allowed the construction of the internal combustion engine; the equations of electromagnetism on which electrical engineering, optics, and modern telecommunication are based, and finally quantum physics, disclosing both the hidden behavior of matter, and the history of the universe as well, from the Big Bang to our times.

But any success is bound to degenerate. According to the French philosopher Edgar Morin [11], one of the greatest living representatives of complex thought, the paradigm of simplification gives birth to a blind intelligence. Assigning importance to simple parts only, always insisting on the same direction, more and more in detail, the risk is ignorance of the complex system of which they are a part. Once the building is dismantled, the possibility of building it again starting from the fundamental bricks cannot be taken for granted. This seems an intuitive concept.

Nevertheless, physics have been dominated for centuries by scholars of elementary particles, interested only in the smallest components of matter, and it was so up to 1972, the year in which Philip Warren Anderson [1], later winner of the Nobel Prize, published in the magazine *Nature* the historical article "More Is Different," in which he maintained the whole is larger than the sum of the parts, and much different from it.

It is true, Anderson writes, that living matter is made of cells, cells are made of molecules, and so on, up to nuclei and quarks, but the knowledge of elementary constituents and the forces acting on each other, in order to reconstruct the properties of matter on a larger scale, is not enough, because we "are met with truly fundamental questions every time we put the parts together in order to form a more complex system and we try and understand the ensuing substantially new behaviors."

The existence of physical laws is amazing, but it's not all. The natural world is ruled both by essential principles, and by the extraordinary organizational principles at the basis of structures that acquire meaning and autonomous life, and self-organize from the bottom until they transcend the parts of which they are made. The liquid state, as an example, is an emerging property that doesn't belong to the single water molecules. Conscience doesn't reside in neurons. Economies, politics, cultures (luckily) are not just the summation of people.

The tension between the two currents of thought (the law of parts or the laws of the collective) is the carrying axis of the process of understanding of the world. If up to a few decades ago we concentrated mostly on the laws of parts, now the second type acquires more importance, as states the Nobel Prize Laureate for physics, Robert Laughlin [2, p. 208]:

> Science has now moved from an Age of Reductionism to an Age of Emergence, a time when the search for ultimate causes of things shifts from the behavior of parts to the behavior of the collective. It is difficult to identify a specific

moment when this transition occurred because it was gradual and somewhat obscured by the persistence of myths, but there can be no doubt that the dominant paradigm now is organizational.

The border of modern science is here. The border of management is here.

TOWARD SELF-ORGANIZATION

The changes that took place in science in the second half of the twentieth century affected management too. A yet bigger contribution came from the spreading perception of how inadequate traditional instruments are to the mutation of times and contexts.

When life was flowing as slowly as a lazy river, considering and therefore managing organizations as simple systems in simple environments was possible. The well-known saying that the famous Ford T-model could be any color as long as it was black is self-explanatory as to the attitude of the industrial Taylor–Ford model in order to face and solve outer and inner complexity.

Rationality and opportunities offered by technology were deeply trusted, organization was seen as the regulating system of a big mechanism where people act mechanically and predictably, workers are totally replaceable, and project and executive activities are completely distinct. Like a machine, organization is planned from above. The only possible alternative for the machine mechanisms is executing the tasks for which they have been designed. Prescribed rules, formalized control, and structured hierarchical authority all aim at simplifying organizational operations and induce simple and well-defined answers.

But today that rhythm has become as pressing as a whirling stream, and we must admit that, although the classic model is still needed, it's not enough. It's perfect for the branches of continuity, but unsuitable for the points of sudden change. A new orientation is gaining ground: energies have to concentrate on flexibility, or on the ability of keeping one's options open. In this scenario, the fundamental hypothesis is that the future is unpredictable, rather than predictable. Simplifying deliberately, with the aim of underlining the differences in basic orientation, we outline a few of the characteristics of the so-called classic model, in comparison with the complex one.

First of all, change can be easily observed on the structural level. The Ford idea of the all-inclusive company, essentially developing along inner lines, has gradually been substituted by less-integrated types of companies, with a reduced hierarchical structure of the decision-making power. The spreading of outsourcing and the success of net-shaped structures for the defense of network organization brought us to the passage from a culture of possession to a culture of protection: integrating all activities in a single vertically organized company changed into using the activities put at everyone's disposal inside a relationship network. The passage has been from mainly formal co-ordination mechanisms to mainly informal ones.

In the classic model, a strategy is first studied on paper: later it's implemented from above, looking for agreements, order, and harmony. Afterward, planning becomes essential: centralized, accurate, detailed, perfect in theory, analytical, and, more often than not, with a faraway deadline. The top management's prevailing role is checking the possible deviations from the plan with an executive orientation, dedicated to the absorption of the deviations, in order for the plan to be brought to conclusion as forecast and the strategy implemented.

The complex model starts from the assumption that exploring strategies available to their own change is more profitable than long-term forecasts and assures development in highly mutable environments. Such strategies may also emerge from the bottom and make allowance for parts of conflict, disorder, or discordance. Planning is decentralized, and especially sensitive toward weak signals coming from both inner or outer peripheral zones, and becomes the construction of scenarios including the overview of all possibilities. Top management's task is to create the conditions to learn and innovate with an exploring attitude, accepting the fact that the direction to choose will be defined only once the conditions arise.

The fundamental difference stands in the basic assumption and in the goal the top management sets up to face complexity. The classic model aims at stability (complexity reduction), the complex one at flexibility (complexity absorption). In order to learn more and extend the understanding of these concepts, please refer to Chapter 4, by Alberto Felice de Toni.

According to Ali Jaafari [3, pp. 47–57], a sudden change as to the model to apply is taking place in project management as well. Referring to Table 5.1, a situation of high complexity can be found both outside the project and, more and more often, within the project itself. If complexity is low, a possibility is not to use any type of model, or to create a very simple

TABLE 5.1

Comparison between Classic and Complex Managerial Models

	Type of Managerial Model	
Characteristics	**Classic**	**Complex**
Structure	Hierarchy	Network
New strategies formulation	Top down, backing agreement, order, harmony	Bottom up, accepting conflicts, disorder, discordance
Planning	Centralized, analytic	Decentralized, by scenarios
Managers' prevailing role	Planning and control	Creation of learning and innovating context
Managers' prevailing trend	Executive	Exploratory

model for the purpose; we're possibly not even facing actual projects. At the first increase in complexity, the association with a bureaucratic model, consisting of checks, proceedings, and administrative processes, a model shared by many public projects, is possible. The next step is the intermediate situation of the regulative model, where even very complex projects might have to be managed, but the environment is either not so complex yet, or can be assumed, shaped, and simplified as such.

Figure 5.1 shows the model we use every day as project managers, trying to reduce complexity by arranging goals and structures suitable for

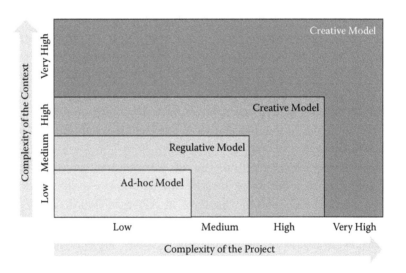

FIGURE 5.1

The two dimensions of complexity. (From Jaafari, A. *Project Management in the Age of Complexity and Change*. Project Management Institute, 2003.)

planning in an orderly way and delivering the project. Today, both in project and in context, we shift more and more toward the regions on the edge of chaos, and greater and greater importance is acquired by the absorption of complexity, thanks to a creative model, based on self-organization, flexibility, and change as an answer to circumstances.

The emergency era, as we were saying, shifted the focus over community laws, that is, in particular over the principle of self-organization, over new properties, structures, and behaviors emerging from the bottom, that are not present at the level of individual parts: this is why we say that the whole is larger than the sum of the parts. In the following I describe the organizational principles and the project management topics born out of the search for analogies with what happens in self-organized physical and biological systems. In the beginning of the chapter I outline the main characteristics of four advanced organizational models:

- Circular organization
- Holographic organization
- Cellular organization
- Holonic organization

For each of them, the following organizational principles are identified [4, pp. 208–234]:

- Organizational principle # 1: Interconnection
- Organizational principle # 2: Redundancy
- Organizational principle # 3: Sharing
- Organizational principle # 4: Reconfiguration

Subsequently, each principle is associated with a project management topic, among those better highlighted in the next chapters of this book. These topics are:

- Project management topic #1: Relationship
- Project management topic #2: Redundancy
- Project management topic #3: Ethics
- Project management topic #4: Governance

As a conclusion, the four recurring topics that, in my opinion, allow the synthesis of the so-called self-organized project management logics (and make sure this expression doesn't turn into an oxymoron) are highlighted.

IDEAS FROM THE "CIRCULAR" ORGANIZATION

Circular organization (see Figure 5.2) has been described at different times by two of the main management experts: Henry Mintzberg and Russell Ackoff. In 1966, Henry Mintzberg [5, pp. 61–67] wrote:

> Try this metaphor. Draw the organization as a circle. In the centre stands central management. And in the outer points stand the people developing, producing and delivering products and services—those operating daily. The latter see very clearly what happens, because they are closer to action. But they have a limited vision, they can only see their small segments. The managers at the centre enjoy a wider vision—all around the circle—but they cannot see clearly because they are far from the action. The trick consists in connecting the two groups. This is why many organizations need informed managers in the centre, people able to look towards the outside then turn and speak with those in the centre. Yes, the people we used to call middle managers, that almost disappeared from our organizations.

Mintzberg suggests trying to eliminate the vertical metaphor, used to represent hierarchy, and substituting it with the circular one, with interconnected centers and peripheries, exploiting the effect of small worlds: small semiautonomous circles inside with closely connected knots, which are linked to other units through structures called hubs, that is, persons/ structures enjoying a high number of links (contacts) with the outside. The organizational principle it concerns is interconnection.

According to Russel Ackoff [6], one of the founders of management science and a great expert on operative research, this represents an evolution in hierarchy, a kind of trade-off between the top-down structure and the

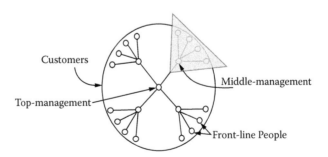

FIGURE 5.2
Characteristics of a circular organization.

bottom-up one (in English in the text) that could be called "democratic hierarchy," where the holder of authority is in his turn submitted to the collective authority of the rest of the group.

The organizational principle of interconnection refers to one of the project management topics that is studied more extensively, the topic of relations. The project team can be considered a semiautonomous circle connected to the semiautonomous circles to which stakeholders belong. The ability of establishing and maintaining relations, both inside and outside the project team, has always been considered a key skill in project managers; in the territory of complexity, where many linear and nonlinear connections exist, and therefore unimaginable effects, both threats or opportunities can arise, therefore understanding their dynamics acquires greater and greater importance.

What we, as project managers, should never forget, is to add the time arrow to our diagrams, to our representations, to our mental schemes, and to considerations. Everything moves in complexity, very small space is actually left for static balance. Goals, behavior, interests, and emotional state in the team members and the stakeholders can change considerably while a project is being carried out. This requires dynamic management of relations.

In Chapter 7, "Stakeholders' Worlds," this requirement is highlighted: "[T]he analysis of the stakeholders' world cannot be limited to an initial snapshot, no matter how realistic, but it shall monitor its evolution over time. Complexity of the organizations indeed calls for an analysis which takes into account relation dynamics (both in individual and group perspective), cultural framework and individual creativity." In that chapter, some useful instruments are also highlighted: for instance, the "social/organizational network analysis" allows informal networks, often overlooked, to be studied as well, and "type-watching" helps approach other people, so that relationship gets better on both sides.

Relations change more and more in the course of the project, and should be considered inside the co-evolutionary dynamics of the context or contexts concerned. Different networks can appear, change, or die, just as in an ecosystem, and project management is required to understand them and act as connector. These networks will be formal ones, delicately balancing with informal networks over the edge of chaos. Therefore, as outlined in Chapter 6, "The Project beyond WBS," a kind of systemic map taking into account the whole dynamics of relations and balances can be very helpful. WBS is fundamental, and should never be given up, but new instruments are called for in order to support the project manager and avoid his facing unexpected events unprepared.

There is such a wealth of informal relations in the instruments we, as project managers, use every day! Up to the present day we simplified, and this helped enormously in our profession. But Chapter 12, "The Value of Redundancy," underlines once more that the fact of encouraging and managing a redundancy of interactions represents one of the possibilities of developing inside the project team the flexibility required to react to unexpected sudden events. It's the law of necessary variety in cybernetics, also known as Ashby's law: to a great variety of outer disturbances, to great complexity, we could say, an equal variety of available answers must correspond, a redundancy of relations, in this case, able to ensure regulation of the system in order to keep within a limited number of states. As semiautonomous circles in a perpetually dynamic relation …

IDEAS FROM THE "HOLOGRAPHIC" ORGANIZATION

The holographic organization (see Figure 5.3) has been suggested by Gareth Morgan [7], the author of the well-known *Images of Organizations*. Organizations should learn something from the self-organizing abilities

FIGURE 5.3
Characteristics of a holographic organization.

of the brain, especially from its redundancy, which makes it similar to a hologram. A hologram is an image where each of its parts contains every piece of information in order to reproduce the whole, so that, should it go to pieces, the whole original shape could be recreated out of each part. Redundancy is precisely the key of evolutionary flexibility, and holographic organization is betting on this principle, as it ensures the possibility of quickly reassuming its form.

There are two ways of planning a redundant system: acting on the redundancy of the parts or on the redundancy of functions. In the first case, somehow inessential parts are added, to take care of control, support, or replacement. This choice increases complication, without nevertheless relevantly affecting the company's adaptability. In the second case, however, this effect is plainly reached. No part is added to the system, but additional functions are assigned to each part, so that the latter can carry out a differentiated range of activities rather than a single specialized work. The newly acquired abilities might not be of immediate use, but prove crucial for survival in the future.

The fundamental point to understand is that the operative effectiveness is not all but, on the contrary, we may have to give up some of it in order to increase strategic options and ensure a timewise extended adaptation ability. A wider range of viewpoints allows more variables and interconnections to be taken into account; therefore the chances of being caught unawares by unexpected events affecting the organization's very life are reduced. The higher cost at the beginning is more than compensated for by the long-term advantages.

In such a system, learning is fundamental: as far as possible, "everyone learns how to do everything." Special attention must then be paid to the border relations between the organizational units and their environment, in order to ensure the required variety discussed in the former paragraph, and make it available to the unity in need. In case of need, it may be possible to move particularly skilled people to different tasks and duties (extra resources). Interchange becomes higher, exactly as in biological systems: first of all in our bodies (we are normally endowed with two eyes, two ears, two lungs, and so on), and also in beehives, ant nests, and termites' nests, all capable of being unexpectedly effective when responding to environmental variations through spontaneous reorganizations. The investment in redundancy, therefore, means in fact an investment in a long-term sustainability of the company.

Back to project management, both the project team and the project ecosystem in the specific context (the project program or portfolio) could be seen as holographic organizations. In both cases, the topic of redundancy appears fundamental for project management too, for the same reasons we just analyzed. In fact in the following pages the group of the Complexnauts dwell on it many times, and dedicate to this area the whole of Chapter 12 ("The Value of Redundancy").

When we plan a project for a customer, we know we could summarize his request with a single word: reduction: reduction of time and costs and unchanged quality. We apply reduction as much as possible, leaving the smallest margin as acknowledgment of the risk implied in the schedule of activities (called contingency in the traditional discipline). The concept of contingency, at close inspection, has little bearing on the topic of redundancy. Redundancy is not an increase in contingency, but means encouraging the wealth of information, of interactions, of abilities and approaches. The change applied is not a small one: are we ready to give some effectiveness and invest in redundancy? The project in the planning phase will probably seem more expensive, frightening both the stakeholders and ourselves. But how much would be gained in survival abilities!

Such wealth assumes a truly priceless value in complexity, where every project can be considered as a single body, hard to be framed precisely inside predesigned categories. Redundancy can be bet on in several ways: for instance, by making the most of informal networks and encouraging the birth of new ritual behaviors inside the group, as is respectively suggested in Chapters 7 and 8, "Stakeholders' Worlds" and "The Propitious Time"; then by keeping track of events, pieces of information, reports, e-mails, and estimates, on one hand, and of emotions in the shape of episodes, dialogues, tales, legends, and words in front of the coffee machine, as in Chapters 6 and 10, "The Project beyond WBS" and "Narrating to Believe," respectively, and looking for balance on the edge of chaos between redundancy (extra resources to be used when needed) and scarcity (small shortages capable of sharpening one's wits), as in Chapter 9, "Leadership and Complexity."

Looking up from one's own single project, and considering the project ecosystem in its context (the company's project portfolio, my projects, and so on) is often enough. Sources of redundancy are probably available, in the form of resources, former experiences, stories, practices, among others. But, again, it's all about giving up the logic of efficiency at all costs: looking up from one's own single project is often the hardest thing to do for a project manager always struggling with expiration dates, requests, and pressure. Nevertheless it's all around us; we only have to look up. As holograms, redundant for evolutionary flexibility …

IDEAS FROM THE "CELLULAR" ORGANIZATION

The cellular organization (see Figure 5.4) has been suggested by Miles and Snow [8]. It is about a mass of organizational units of variable size, made up of 10–12 professionals up to a maximum of 150 (generally the average is about 15 members). Each cell chooses among its members a project manager, not as a hierarchical boss, but rather as a co-ordinator. The units can assume the shape of centers of economic responsibility (expenses center, cost, revenue, profit, and investment centers), with full decisional autonomy, and are responsible for the results and the effectiveness of processes and are encouraged to enter into alliances inside or outside the organization. This

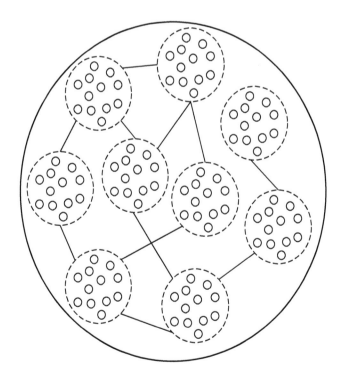

FIGURE 5.4
Characteristics of a cellular organization.

way, everyone can do everything, and acquire an exceptional capability of connecting quickly with actors all along the value-making chain.

The comparison with the human body comes immediately. Each of our cells is endowed with the ability of acting autonomously, but complex human behavior is only made possible by their joint action. Therefore, teams can be self-managed and have entrepreneurial responsibility, and still find their full meaning when sharing with other teams.

To allow synchronic movement, cells are called to share on a first level some procedures or shared services, and on a second level a set of values, a shared culture, a vision, thus safeguarding the identity of the system. On the first level we consider the so-called shared services, specific entities inside or outside the borders of the company used to supply services to the cells.

The second, or social, level (of values, of a shared culture) and the strategic level (of the vision) concern sharing. On the social level, this represents the answer to the innate human need for acceptance and active participation in a group. On the strategic level, the power of vision at the corporate level corresponds to the power of myth at the community level: as myth is the vehicle of change from one social state to another, so does vision lead the company group toward the goal, gives the route of march, sets the borders of inclusion, and supports the feeling of belonging. Vision, like myth, is endowed with a founding value for the group's origin and purpose, and the group becomes the ideal domain connoting each person's development path.

Project teams and members of a single team, as well, can be considered cells, acting autonomously but yet able to produce satisfying results only thanks to their joint action. In project management, sharing on the second level can still be associated with the topic of fundamental values, or, more generally speaking, to the topic of ethics. A proper foundation of ethics requires a standard of values as model for all goals and actions, applicable both in our day-to-day choices and in emergencies. PMI® actually supplies the "ethical and professional behavior code," and suggests the respect and sharing of four main values, singled out by the global community of project managers: responsibility, respect, fairness, and integrity.

The greater the unpredictability, the greater is the importance of ethical behavior: when everything outside is unpredictable, more and more so, our duty as human beings, and in our case as project managers, is to increase predictability, growing predictability, in order to keep the delicate balance on the edge of chaos. Ethics are required for the addition of predictability.

Complexity is the mother of freedom, making us active subjects of unimaginable realities created by ourselves, if it is true that a small cause, acting in the right direction, can lead to totally unpredictable effects. We could potentially change the world, because everything is nonlinearly connected. Faced with new and unexpected situations, we can use a great potential and direct it toward creation or destruction. This is the ethical viewpoint of complexity. From this viewpoint, ethical behavior inside a project team is aimed at creating the environment needed to allow each member of the team, in unpredictable circumstances, to give a consistent, autonomous, and predictable answer.

In several chapters of the book ethics are mentioned in this sense. In Chapter 9, "Leadership and Complexity," ethical behavior is considered a determinant in managing unstable situations. The project manager should be close to team members in critical moments through proactive listening and try at all times to remove inner disagreement and promote reciprocal respect.

In Chapter 11, "Risk and Complexity" (in the section "Ethics for Harmony"), the authors go further, by maintaining that, more than advanced techniques of risk management, what really matters is sharing a system of values inside the project. This allows each member to act and react naturally and consciously to unexpected events in the project. Thus is each person's compliance built, in the form of adequacy to the environment (to other people, to the context, etc.), and trust accepted as an instrument of relation then more trust deliberately generated, through comparison and humility, that is, once more, through the inclination to listening.

We will all be asked, more and more often, to behave ethically, if it were only for the positive results of project management, particularly in critical moments. The fundamental values should be shared inside the group, and a greater strength in this sense will be acquired by storytelling, used as a cement in order to create a shared environment (please see Chapter 10, "Narrating to Believe"). The complication we are used to can be condensed in a system of equations, or in our specific case, in deterministic instruments such as Gantt, WBS, Pert, CPM, and so on. Complexity, on the contrary, needs to be told. As independent cells, but sharing. …

IDEAS FROM THE "HOLONIC" ORGANIZATION

The model of holonic organization (see Figure 5.5) suggested by McHugh et al. [9] was born with the aim of determining the organizational principles of just-in-time. This is a special evolution of the classic network

FIGURE 5.5
Characteristics of a holonic organization.

enterprise formed by a system of companies or independent units, able, in case of need, to integrate in freedom in order to answer particular business needs, and put into action, in doing so, a virtual subject. The organizational principle of reference is reconfiguration: new configurations, capable of evolving together with the environment, emerge all the time.

In order to understand better, let's consider the field of cinema: every new movie actually is a new network, in which scriptwriters, technicians, director, and the actors get together and form a virtual enterprise (a project) that will dissolve at the end of shooting. Such dynamics can be traced in many fields on the border of technology (biotechnology, nanotechnology, etc.), where the main issue is finding the required resources (researchers, patents, niche manufacturers) anywhere in the world. The holonic enterprise is therefore a network system able to mobilize decentralized intelligence and its creativity, arranging the required forms in order for a project, born at any point of the network, to quickly and effectively arrange around itself, then integrate in a system, all resources and intelligence spread around the network as a whole [10].

Any of the potential decision-making centers, or organizations belonging to the network, can assume command, or, better still, the role of catalyst, at any moment, according to the various stimuli coming from the reference environment. There's no lack of leaders, but they vary according to the situations and to the emerging business opportunity around which the virtual enterprise is getting organized.

The value chain is not definite but changeable. The skills that are today's core of added value offered to the final customer, can tomorrow act as a mere commodity for another product on a market completely different from ours. And the latter will not automatically produce a reduced value for the company, if made available on the Net. In any specific situation what allows the best pursuit of the present business opportunity emerges. The parts are horizontally connected, and associated with a shared information system and shared values.

For many people, being part of a holonic network assuming a new shape all the time, so as to adopt the role of promoter and co-ordinator on a certain market, and simply the role of supplier on another, seems a sustainable solution in order to seize business opportunities, that is, the opportunities for profit, anywhere in today's global economy.

Back to project management, the recurring topic linked to reconfiguration is the topic of project governance (project driving). Together with the topic of redundancy, the latter seems the topic affecting the discipline with greater strength. It has been discussed in several parts of the book, but it is especially defined and developed in Chapter 6, "The Project beyond WBS," where it is said that "the idea of monitoring and controlling should develop into a concept of project driving, that is, the set of actions which can be used by project managers to make the most out of the conditions that emerge as the project evolves in order to be able to define the most appropriate strategy (at a given moment) to achieve the expected goal." Francesco Varanini explains, that such is the helmsman's steering: he knows the port of destination, and, by subsequent adjustments, maintains the course.

The project being by definition a temporary social organization dynamically evolving in time, recurring effectively to planning and control is quite difficult. We are in danger of ending up not managing a project for the realization of something, but the project of managing variances to the initial program. Sometimes our efforts are directed more to project management in itself, to techniques, to the respect of engagements with customers, than to the actual output.

Uncertainties and unforeseen events of all kinds encountered by a project on its way can be considered as unexpected opportunities to be caught and not threats to be avoided or hidden. To travel one has to go and preconditions to turn events in our favor must be created. The more we are able to understand weak signals, to go along with transformation, to create preconditions (for instance, acting on other topics: relations, redundancy, ethics), the quicker we will reorient the project in case of need. A kind of co-evolution of the project, applying the distinctive method of Chinese vision: evaluate the context and exploit its potential (evaluation and exploitation).

The governance of the project also includes enhancing the concept of time, as explained in Chapter 8, "The Propitious Time." Under the logic of planning and control, time is a linear sequence of activities and related events, with an objective, predictable, and measurable duration, a precise start, and an exact end: here it is, the artificial time of the Gantt diagram. But today a qualitative, rather than quantitative time, a subjective, rather than objective time, a time connected to human activities, to opportunities, to change, to dynamics, and not to measurement, acquires more and more importance. At the propitious moment, not at the moment Gantt prescribes, the potential of the situation can be exploited and the absorption of complexity can be attempted.

Project governance is further discussed in most chapters. In Chapter 9, "Leadership and Complexity," the new leadership required from project managers is synthesized by the motto, "We are all ready for everything any time." Here our main point is that project governance is much more effective emerging from the bottom, from the contributions of all single components, than if implemented from the top. Everyone is called to feel part of the project, a leader, and personally committed. In Chapter 11, "Risk and Complexity," on the other hand, the highlight is set on the need for a project manager to manage risk dynamically in the points of discontinuity; the key to success is the creation beforehand of a shared wisdom, to be used later when needed. As are holons, they are ready at all times to take another shape.

THE TOPICS OF SELF-ORGANIZED PROJECT MANAGEMENT

The four topics of self-organized project-management, selected according to the organizational principles the more advanced models refer to, have been associated with the topics of the research developed by the Complexnauts (see Table 5.2). As seen in the text, self-organized project management is discussed from different viewpoints in the various chapters of this book. In Table 5.3 the topics of self-organized project management with the specific subjects discussed are synthesized.

TABLE 5.2

Topics of Self-Organized Project Management Discussed in This Book

| Topics of the Research | Auto-Organized Project Management | | | |
	Relations	Redundancy	Ethics	Governance
The project beyond WBS	☑			☑
Stakeholders' worlds	☑	☑		
The propitious time		☑		☑
Leadership and complexity		☑	☑	☑
Narrating to believe		☑	☑	
Risk and complexity			☑	☑
The value of redundancy	☑	☑		

TABLE 5.3

Topics of Self-Organized Project Management Discussed in This Book

Topics	Specific Subjects
1. Relations	Dynamic balance between formal and informal networks
	Integration of traditional instruments to trace all relations
	Redundancy of interactions
2. Redundancy	Enhancement of the value of informal networks
	Appearance of new forms of ritual behavior in the group
	Tracking of all events, pieces of information, emotions, stories
	Dynamic balance between redundancy and shortage
3. Ethics	Elimination of the frictions inside the group
	Individual spontaneous reaction to unexpected events (compliance with the environment)
4. Governance (project driving)	Overcoming the planning and control approach
	Exploitation of the propitious moment
	Leadership emerging from the bottom up
	Dynamic management

CONCLUSIONS

The expression "self-organized project management" may be an oxymoron. Here it has been used as provocation, in order to ask questions, to try to go beyond it. Of course, the small certitudes we cling to day by day will have to be let go. Simplifying is rather reassuring, but, because it's becoming dangerous as well, some of our security should be set aside. Let's give up some security and let's start our journey.

The best aspect in complexity is its insinuating doubts at all times; settling down is impossible. With this statement, the invitation is to go on with the research, to look further into the matter, to start new journeys toward unknown places where order and disorder coexist as day and night do at sunset.

What I wish you to do is to go to work tomorrow and, as every morning, switch your PC on, give a look at your Gantt and at your WBS, check your project charter once more so that nothing slips your mind, keep track of the main risks and of who is managing them, organize an alignment meeting with your project team in the morning and another with the customer in the afternoon, reply to a customer's e-mail where he is pressing for a deliverable that won't actually be ready before next week, and send a high priority e-mail to your project team in order to have this deliverable ready before this term. But I also hope that, in a quiet moment, maybe in front of your coffee, you would start wondering. Something is amiss, in this project. Your Gantt is perfectly all right; the WBS is perfectly all right; the corporate procedures have to be respected, no doubt. But this project is neither the project you did before, nor the future one. It's not what it is today nor what it is going to be tomorrow. This one, at the moment, is "this very project, right now." A unique moment.

REFERENCES

1. Anderson, P.W. (1972). More is different. *Science*, 177, 393–396.
2. Laughlin, R. (2005). *A Different Universe: Reinventing Physics from the Bottom Down*. New York: Basic.
3. Jaafari, A. (2003). *Project Management in the Age of Complexity and Change. Project Management Journal*, 34(4): 47–57.
4. De Toni, A.F., Comello, L., and Ioan L. (2011). *Auto-organizzazioni. Il mistero dell'emergenza dal basso nei sistemi fisici, biologici e sociali*. Venice: Marsilio.
5. Mintzberg, H. (1996). Musings on management. *Harvard Business Review*, 74: 4, 61–67.
6. Ackoff, R.L. (1989). The circular organization: An update. *Academy of Management Executive*, 3: 1, 11–16.
7. Morgan, G. (1986). *Images of Organizations*. London: Sage.
8. Miles, R.E. and Snow, C.C. (1995). The new network firm: A spherical structure built on a human investment philosophy. *Organizational Dynamics*, 23: 4, 5–18.
9. McHugh, P., Merli, G., and William, A.W. (1995). *Beyond Business Process Reengineering: Towards the Holonic*. Chichester, UK: Wiley.
10. Caruso, E. (2000). *L'eccellenza nelle imprese*. Milan: Franco Angeli.
11. Morin, E. (1990). *Introduction à la pensée complexe*. Paris: Seuil.

6

The Project beyond WBS

Livio Paradiso and Michela Ruffa

CONTENTS

Marco Polo describes a bridge, stone by stone. "But which is the stone that supports the bridge?" Kublai Khan asks.

"The bridge is not supported by one stone or another," Marco answers, "but by the line of the arch that they form."

Kublai Khan remains silent, reflecting. Then he adds: "Why do you speak of the stones? It is only the arch that matters to me."

Polo answers: "Without stones there is no arch."

The Invisible Cities **by Italo Calvino**

PROJECT MANAGERS THROWN INTO THE FRAY

Scenario #1

During the monthly progress meeting, I watched the customer, reassured by the fact that the project was progressing according to the plan that we had defined and approved. He exclusively focused on verifying dates, deadlines, costs, and progress. Everything was going according to his expectations.

As a project manager, I was sure I was doing a good job, and a satisfied customer was clear evidence of that. But I could feel a slight sense of anxiety, similar to the emptiness that grabs you when you realize that you might have overlooked something and that reality might be different than you thought, different from the way you perceive it and no longer under your full control.

Why was that? Why did it suddenly show up, right at that time, when it was clear that I had done my best to accomplish my tasks with all of the official blessings? I started to rack my brain to spot what could have gone wrong, where the threat could come from, but from my list of manageable variables I only got positive feedback. What, who, when, how much, risks, assumptions, stakeholders, communications: everything had been defined carefully and was being checked as well. The team was motivated and cohesive. So why was that?

Scenario #2

Team member: "Listen, I should tell you something; it's quite urgent."
Project manager: "I'm all ears."
Team member: "You know, the specifications we froze some days ago …"
Project manager: "Yes?!?"
Team member: "I spoke with one of the key users who told me that yesterday they found an additional feature which could be really useful for them, and we could incorporate that feature with a new technology which maybe will bring us some cost savings."
Project manager: "No way! It's been so hard to get to a frozen scope for this project; we cannot afford listening to whoever wakes up with a new feature in mind that could be crucial for the project. … Unless they are willing to consider it as a scope change!"

Scenario #3

Team member: "Listen …"
Project manager: "Yes …"
Team member: "The new development manager just mentioned a design methodology which could save us a lot of testing time."

Project manager: "That's interesting. ... But we should modify the whole development plan. And who is going to explain it to the customer? And what about risks? No, let's leave it as it is; let's go on as planned: let's keep it in mind for the next project."

Scenario #4

I have a final WBS, I have assigned the roles, and I can rely on a detailed control system. I know I am compliant with all the required steps. The project is progressing as planned. But am I achieving the final objectives? The customer, my main stakeholder, is not happy with the results. Was he expecting something different? The contract is honored, but the final result seems far away. In order to pursue the objective, am I doing the right things, or should I do something different?

PROJECT MANAGER: POWER OR FRUSTRATION?

Whoever has to deal with projects, no matter which kind, will sooner or later realize that the project environment is a harsh one: no matter the role, being it more or less operational, being a main player or a supporting figure, and loaded with responsibility or not. Everybody gets to the point when it is clear that the project context is by no means that easy and malleable reality that we have been led to believe: complex and articulated, but nevertheless manageable by strictly and scientifically applying the processes and the tools provided by the PMBoK®.

Every project manager knows that this profession calls for a good amount of conscience (from the Latin *cum-scire*, "know together"), a considerable dose of consciousness (i.e., the perception and cognitive reaction that an animal exhibits when a given condition or event happens), deep concentration, and the right mix of specific skills, which have been provided by both theory and practice.

The PMBoK, which is the methodological and cultural reference for so many project professionals, presents a project manager who owns the ultimate power and who possesses the right control levers to make the project mechanisms perform at their best, by strictly adhering to the provided directions and correctly applying both the approaches and the tools provided by the common reference.

Experience tells us (sometimes in a tough and painful way) that reality can be quite different from that, and that sometimes conscience, consciousness, and competence are not enough for the project manager to perform to the best of his ability. This is not due to the fact that "the other stakeholders" do not let us act as we planned, neither it is due to "adversity of destiny": rather, it is the nature of the project itself preventing this. Even if the project manager had more power, or better skills, the situation would not improve, because there are some factors (both internal and external to the project) that are just unpredictably *a priori*.

The paradigm of "Project Manager/God" (see Figure 6.1) suggested by Francesco Varanini shows how the project manager experiences, during the project life cycle, a series of conditions and situations that are poles apart.

At times we feel as powerful as masters of the world; at other times we feel only the obstacles and the difficulties, as a burden on our shoulders. Sometimes we experience actions and choices we feel are driving efforts in the right direction; sometimes the same actions and choices just look totally useless and unrelated to the result we want to achieve.

When we realize that it is impossible to influence the fate of the project, frustration arises, because it seems as if we do not have control. Even when we feel like God, we feel the burden of the unpredictability of the causes that led to that condition, which can be lost any time. We feel both

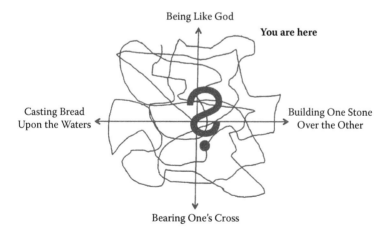

FIGURE 6.1
Conditions and situations experienced by a project manager.

inadequate and doomed to chase after the project tasks, rather than to determine their development, as if the project lived its own life, able to shun any chance of control.

FORECAST AT ANY COST: THE MYTH OF THE BIG BROTHER

One of the basic principles of classic project management is that a project can be successfully and efficiently controlled by:

- Forecasting in advance what is going to happen and defining in detail the actions to be implemented (planning)
- Being able to verify at every single moment that what is actually going on is consistent with the forecast (control)

This is the idea underlying the process group division presented by the PMBoK, which defines (in addition to "initiating" and "closing") "planning," "executing," and "monitoring and controlling" [1, pp. 39–65].

Planning and controlling are indeed decisive for project success under the assumption that each and every fundamental element of the project is predictable, and that the correct set of key performance indicators (KPIs) is chosen. The more the project evolution sticks to its initial definition, the better the project performs; calculation of both effectiveness and efficiency based on the earned value method is the clearest evidence of this concept. Performances are usually measured as the capability of replicating in the project reality whatever has been planned in advance. Uncontrolled variables and risks are relegated to recovery plans and contingency reserves, both of which are aimed at being minimized and neutralized. Even the risk management tools described by the PMBoK are based on the capability of forecasting potential future scenarios, in order to reduce resulting interference and pursue the completion of the established plan to scope.

But indeed, our experience leads us to consider that sometimes planning and control are not the keys to the success of our project. Planning, monitoring, earned value, and risk analysis are basic, useful, and indispensable tools. They represent a necessary but not always sufficient condition for project success whenever project complexity dynamics arise. First of all, this is because the number of variables that influence and interact

FIGURE 6.2
Project plan syndrome.

with the life cycle of the project is constantly expanding and the outcome is less and less predictable.

Secondly, the instability of the context undermines the possibility of making a detailed and affordable forecast. By *context* we mean the project's operating environment, and also the environment surrounding the project (the business scenario composed of all the organizational entities involved in the making of the project), or overall project acting environment, whose scope, nowadays, is getting more and more multidisciplinary and globalized.

In our experience, few projects started with both specifications and objectives that were clearly defined, planned, and successfully controlled without substantial changes along the way. And the hunt for the "still-to-be-subdued" variable might easily remind us of Don Quixote's attack on windmills (see Figure 6.2).

DIVIDE AND CONQUER: THE LIMITS OF THE ANALYTICAL APPROACH

Another fundamental principle of the project management body of knowledge is that the "whole" is defined as the algebraic sum of all the components. Knowing and controlling these components is sufficient to manage the whole. Therefore, when starting a project, it is worth decomposing the whole, aimed at focusing on its most basic components, inasmuch as:

- They are easier to manage, both from a planning and a controlling perspective.
- Their management leads to the management of the project as a whole and therefore paves the way to the success of the project.

This principle is the basis for the use of hierarchical structures in the definition of the project plan, first of all the work breakdown structure (WBS), which is also the tool used to define the scope of the project. The logic behind hierarchy, combined with the rigor applied in the definition of the levels of synthesis and detail, shows the influence of both the analytical approach and reductionism, that has characterized traditional science in the last few centuries.

Practical experience, however, clearly shows that a strict analytical approach may limit the completeness and accuracy of the overall picture, because important facets exist that can neither be modeled nor described within this decomposition schema. For instance, each project features a whole set of factors that are intrinsically both "intangible" and difficult to explore, but very influential to the final outcome: that is, the feelings, expectations, and points of view of the stakeholders. There is a "a world" surrounding the WBS that is not represented by the WBS itself, and not included in the scope and management tools of the project managers, which are based on the WBS alone.

Moreover, hierarchical structures can be used to represent the components of the represented reality, but not the reciprocal influence relationship among the components themselves: WBS identifies the activity to be carried out, but not the potentially existing connections among the activities, not only the ones related to operations, but especially the ones related to the organization, the opportunities, convenience, and so on. These connections, which can determine an important reciprocal influence among the involved activities, may cause an activity to disappear and another one to change its features completely. Neglecting these connections may jeopardize the correctness of the model defined by the project manager by means of the WBS.

Even the PMBoK reference model recognizes these kinds of connections throughout all the different identified processes: analysis of the project context, of its environment, of the general cultural factors, and of the stakeholders. These aspects are considered of high importance, but on the other side they are embedded in a static model exclusively based on the concept of linear forecast, planning, and control. Again, the WBS acts as a filter focusing on what has to be done even if it is rectified by taking into account environmental factors. By the WBS tracking and focusing on the work to be done within the project, the project manager may lose sight of other important project elements, because the WBS itself does not integrate the tracking of the links and item interdependence developed by the project team to write the WBS elements down. Furthermore, according to the traditional model, the project manager must focus on the planning and control elements of the project.

Whenever connections, environmental factors, and instability become so relevant and dynamic that they cannot be neglected, nor frozen by a pre-emptive analysis, the WBS can act as a screening and limiting filter for project reality, which might be misleading instead of being a firm guiding light. Whatever has been neglected might be more influential than the aspects factored in the model (see Figure 6.3), and the experience proves that it is happening more and more often.

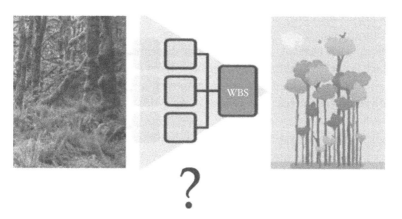

FIGURE 6.3
Limits of a reductionist vision.

THE PROJECT: SET OR SYSTEM?

The topics discussed in the previous paragraphs provide a good reason to look for a different way of approaching a project: our profession encounters several challenges and we do not want to face them while suffering from a condition of inadequacy; therefore we wondered whether it is possible to change and integrate the assumptions that constitute the solid base of the concept of project management. The aim of this exploration is to identify a way of performing as a professional project manager, minimizing frustration, and maximizing the success rate of projects.

We started from the assumption that the project, in general, creates "something that did not exist before," and introduces novelty and change, which derives from the convergence of the "visions" of its stakeholders. It deserves to be considered referring to these important and fundamental features. Complexity theory gave us a hand.

THE PROJECT AS A COMPLEX ADAPTIVE SYSTEM

To begin with, complexity theory introduces the concept of the complex adaptive system (CAS),

> [A] system composed of a great number of elements, both in type and in quantity, which are characterized by non linear connections (that is, the reciprocal influences of action and reaction existing in a dynamic condition); such elements and connections determine an overall behaviour which is not equal to the sum of the behaviours of the single elements, but it also depends on their interaction; moreover, it is able to react and adapt with respect to the reference context [2, pp. 7–13].

Complex adaptive systems are open systems from an energy utilization point of view (exchanging energy with the surrounding context), but closed systems from an organizational point of view: that is, they "can preserve their individual characteristics unmodified, so that it is always possible to identify them despite the structural changes due to their organization" [2, pp. 89–90].

CAS are always looking for the edge of chaos, a condition of unstable equilibrium at the boundary between order and disorder, which is the necessary condition to determine those "discontinuities" that produce innovation and, therefore, evolution. That is why a CAS can achieve self-organization. "Self-organization is the spontaneous emergence of order, of new structures and new forms of behaviour in systems which are thermodynamically open but organizationally closed, at a state far from equilibrium" [2, p. 108].

Both the nonlinear nature of connections and the search for the edge of chaos are the reasons why the CAS are no longer associated with the concept of forecasting, but possibility space is used instead, the space where "everything is possible, but only if something happens" [2, p. 169]. Therefore it is "impossible to predict with certainty the future state of a complex system, even if its obvious possible states can be forecasted in general."

Which definitions could describe better than that the real nature of a project? Which statements could better explain the behaviors that we have identified as "criticalities" (both in the previous paragraphs and in our personal experience), which indeed are logical and natural consequences of a nature that is both complex and adaptive?

Isn't it true that the project is composed of a vast number of elements, both in type and in quantity, that are characterized by nonlinear connections? The relationships that develop among the components of the project (for instance, the relationships among the stakeholders, the impact of an unpredicted event, the consequences of a decision) are indeed nonlinear relationships; that is, they are interactions which create reciprocal influences between the components (action and reaction), not only a unidirectional conditioning. Nonlinear relationships produce a dynamic condition of both the relationship itself and the system as a whole. Within a project, it is scarcely ever possible to determine neither the cause nor the effect of an event, because the entities involved, regardless of the number of variables that came into play, act, undergo, and react at the same time, leading to an overall state of the system which can be attributed to neither the cause nor to the consequence, nor to any form of linear combination of the two conditions.

And what about the self-organizing capability of the project? By definition, it is an entity that generates change in its entirety (because it produces something that did not exist before), standing outside the context while co-evolving together with it.

If the project is really a complex adaptive system, we have to view it through different lenses than the ones we have used thus far as project management professionals: lenses that widen our perspective, making our attitude more appropriate, more comprehensive, and less exclusive.

Complexity must be undertaken using a systemic approach, not an analytical one, considering reality as a whole, instead of the sum of single parts. It is mandatory to embrace a wider view together with the relationships between the weak and the strong elements of which it is composed.

Complex behavior cannot be predetermined a priori because it is self-determined; therefore it can only be understood while it is emerging, while it becomes evident. Change is embedded inside the nature of complex adaptive systems, it represents the essential condition for progress and evolution, and it cannot be eliminated, on penalty of the death of the system itself. Therefore change, in terms of potential evolution and co-evolution, is an opportunity to pursue, a value to look for, and not a criticality or a disturbance that should be minimized, if not altogether eliminated.

When complexity is considered, the power of connections is no longer a problem (as unfortunately it is often perceived), but rather an advantage, because it is an invaluable tool to spot in advance the feeble signs announcing the arrival of an "emergence." Complexity is the dimension of both the possible and the suboptimal. No unique and certain solution to the problem exists: rather, there are countless opportunities to achieve the results through areas of "next-to-optimal solutions," that represent the best path to the final objective, whenever the predefined paths are not effective. The only way to allow possibility to express itself is to drive and govern complexity, rather than strictly control it.

THE LIMITS OF TODAY'S PROJECT MANAGEMENT

When analyzing the professional tools employed by project managers from the complexity point of view, the reasons for the criticality associated with their use are self-evident and have been described in the initial paragraphs. In fact these tools are a perfect fit to handle complicated situations, although they are inadequate to handle complex situations. Complicated problems, which are composed of elements linked by linear relationships, can be solved by an analytical approach, typical of classic models.

The classic approach simplifies reality by modeling it in a linear fashion. A problem can be decomposed as *n* subproblems which are then reassembled (summed) to get back to the initial problem. Solving each subproblem is equivalent to solving the whole problem. No matter how many components the project has, the overall result can be found by solving each component separately. The whole and its components are related by processes that are totally reversible. Complex situations that are defined by components whose relationships are nonlinear call for a global (or systemic) approach.

> It is not possible to understand the plot looking at the single components, it is necessary to use a synthesis approach, a holistic systems point of view. A piece of cloth can be decomposed thread by thread, and each thread can be analyzed: but the sum of these analyses provides no information on the weave of the whole piece of cloth. An analytic understanding of each and every detail of a phenomenon is not possible; the whole system, considered as an indivisible whole, should be understood instead. [2, p. 14]

An inextricable weave perfectly depicts the nonlinear principle. Let's consider, for instance, an unpredictable event that radically changes the framework which generated the event itself. Once it happens, it is no longer possible to get back to the initial state (it is the concept of bifurcation of complexity theory), but the issue must be solved anyway: either we change the project course to get to our objective or we succumb by stubbornly executing the original plan.

It is evident that the "pillars" of the best practices of project management currently used (such as the concept of project scope, the work breakdown structure, the baseline, the idea of planning and control, the meaning of project performance) have been devised for a world where the analytical approach is always successful, where forecasts are always right, where the control (defined as verification of the coherence between the planned and actual state of all the subcomponents of the project) is the means of achieving the overall result. In such a context, change is a disturbance to be minimized, if not altogether eliminated.

But whenever the project features are intrinsically different, which can only be explained in the light of complexity, the project management tools must be redesigned, at least partially, and integrated with innovative ones in order to let the project manager act in a way that is more in line with the new project scenarios.

This approach matches the desire to widen our horizon and to integrate "traditional" tools with others more suitable to deal with different reference paradigms, according to the "inclusive" logic—rather than the "exclusive" one—which is typical of the complex world (AND culture rather than OR culture).

FROM PLANNING TO PROJECT BUILDING, FROM MONITORING AND CONTROLLING TO PROJECT DRIVING

We believe that when project management is considered in a context of complexity it is mandatory to develop the concepts of planning, and monitoring and controlling, whose limits have already been highlighted, into something more coherent with the features of a complex project.

It is evident that both the need of defining the way to implement the project and the need to guarantee a successful result as much as possible are still valid preconditions even in a complex context. But it is necessary to understand how to conciliate these needs with the lack of precise forecasting and with the self-organized nature of the project, which both characterize a complex context.

In our opinion, it is necessary to further develop the concept of planning, typical of the "traditional" approach (i.e., the creation of a deterministic model that defines the way the project should be carried out a priori) into a concept of project building, that is, the set of steps necessary to build the project without rigidly defining contents and modalities, so that it is always possible to accept and exploit both the emergence (literally meaning a condition that "emerges" as a consequence of the self-organized and adaptive nature of the project) and the change. Bearing this idea in mind, the main effort lies in the clear definition of both the objectives and the guidelines, and not of the exact boundaries and details.

Project building should focus on obtaining the overall picture of the project to be seen from any angle, including strong and weak facets, tangible and intangible results, achieving the convergence of both the ideas and the visions of all the involved stakeholders, neither implying nor pursuing the steadiness, focusing on both the present and the actors defining it (we, here, now).

Similarly, the idea of monitoring and controlling should develop into a concept of project driving, of project governance: that is, the set of actions which can be used by the project manager to make the most out of the conditions that emerge as the project evolves, in order to be able to define the most appropriate strategy (for a given moment) required to achieve the final goal. These are actions focused on the objective and on the present; there is no interest in analyzing the past to forecast the future; actions that consider a systemic view of the project, ignoring useless fine details;

actions that sense any "feeble sign of emergence" while taking advantage of any opportunity, always.

We believe that the implementation of this new paradigm can provide the answer to all those questions and doubts related to the scenarios presented at the beginning of this chapter. What once seemed to be an anomaly to resolve is now a typical feature of a project. The new paradigm can also fully explain a full set of situations: the fear of overlooking something; the frustration which derives, as Bice Dellarciprete told the author, "from the expensive deception of planning," the disorientation of the project manager obliged to cope with whimsical conditions. It is important to understand that we need to either overcome or live with these situations, and even learn how to turn them into an advantage for us and for our project.

PROJECT BUILDING

Organizational Scope and Project Manifesto

A first step to give substance to the idea of project building is to think about the meaning of both scope and baseline in the new paradigm. According to the PMBoK, *project scope* is defined as "the work that must be performed to deliver a product, service, or result with the specified features and functions," and *baseline* is defined as "an approved plan for a project, plus or minus approved changes. It is compared to actual performance to determine if performance is within acceptable variance thresholds."

In a complex context, do these definitions still make sense? If they do, what is the actual meaning of scope and baseline? Without a doubt, the definition of scope referred to a project seen as a complex system no longer represents the concept of a "limited boundary" to the project.

Within a complex system, which is open to its context from an energy utilization standpoint, it is neither significant nor useful to differentiate between what is inside and what is outside (in-scope and out-of-scope actions), inasmuch as they both influence the evolution of the system and therefore they both have the same value and should both be taken into account (it could be said that no inside exists without the outside and vice versa). Furthermore, anything could be either inside or outside, depending on the evolutionary opportunity; there is no inside and outside a priori.

Whatever is inside at a given time could be outside later on, depending on the needs of the objective to pursue, which is the only thing that matters.

Of course, our projects absolutely need some definition of a bounded scope to use as a reference for their evolution: without a boundary that separates the project from the rest of the context, the project would not exist. A complex system is open to the surrounding environment from the "energy" exchange point of view (information, activities, resources, etc.), but it is closed from the organizational point of view, which means that it is able to organize itself autonomously and to keep its original identity over time, keeping itself apart from whatever is "the other."

In our projects, we can imagine the boundary is not determined by the limit between inside and outside, but rather by the criteria that make the project work and evolve with respect to the reference context, those criteria which tell it apart from the "nonproject" and identify its peculiar nature. It is therefore possible to introduce the concept of organizational scope of the project, defined as "the set of criteria that the project decides to apply for its own functioning" (see Figure 6.4). Organizational scope replaces the forecast of "what has to be done" with the definition of the game rules that are used to decide how to do the right thing at the right time, depending on the contingent situation, which cannot be determined a priori.

FIGURE 6.4
Functioning as the criterion to tell "project" and "nonproject" apart.

Just as in a football match: nobody can possibly define in advance the sequence of actions to be played, therefore rules are set, resources are defined (the players), the objective is clear (to win by kicking the ball throughout the other's goal) and the team decides the strategy as the game evolves.

The set of game rules "emerges" out of the convergence of the points of view of all the involved stakeholders, who contribute to the definition of the organizational scope as well as to the creation and evolution of the emerging project charter (see Chapter 7). The project manager is responsible for collecting all the different ideas expressed by each stakeholder, each one biased by his or her own point of view, focusing them toward a vision that is shared as much as possible, according to the suboptimality principle which characterizes complexity and represents, in any case, the best possible solution.

The development of the organizational scope is so similar to the one of the emerging project charter that it can be considered both pertinent and contemporary to it. By taking the opportunity of using the synthetic and systemic approach, and not the analytical one, the emerging project charter can seamlessly encompass the organizational scope. In essence, the traditional distinction between the project charter (the document containing both the main topics and project constraints) and project scope (the set of contents and the course through which the project is accomplished) is not defined in our context in such a way that the two concepts can be merged into a single dynamic representation included in the emerging project charter.

Criteria defining organizational scope can include decision-making strategies, operational approaches, organizational and managerial choices, and behavioral rules, whatever is deemed useful for a smooth evolution of the project. They may also include general strategies pertaining both to the activity to be carried out (having a clear idea of the course to follow acts as a stimulus to making the project work) and the fundamental deliverables. The typical details of the traditional definition of scope are not used, even when they are available, because they are not necessary to identify the boundaries of the project, as explained above.

The organizational scope is no longer an expression of the project contents, therefore it has also lost its role of peculiar dimension, together with schedule and cost, included in the traditional "triple constraint" set related to project objective. And of course it is no longer a binding element for the contract, whose modification triggers the "loved or hated, depending on the point of view," change control process. Organizational scope can be modified as many times as needed and is appropriate, according to the co-evolution and adaptation course of the project. Of course, the definition of the organizational scope of a project is based on project objectives even in a complex context, inasmuch as such objectives justify the existence of the project and represent its ultimate goal.

As already mentioned, traditional project management represents project objectives in terms of content (scope), time, and cost, which together are also known as the triple constraint. In the traditional approach, the way to achieve the expected result is to follow the project plan, which models and embeds the triple constraint; to be more precise, the baseline has to be followed: that is, a specific version of the plan which sets the reference for the development of the project, formally agreed upon and shared by all the stakeholders.

In a complex system, it is meaningless to talk about an objective exclusively in terms of what has to be done in the given timeframe at a given cost, because the activities to be carried out will be identifiable only at the right time, as the project is evolving. Moreover, complexity theory states that a complete forecast is impossible while an unbounded possibility space exists. Therefore, should the objective be defined using the triple constraint method, it would be neither significant nor representative of the actual expected result.

For the same reason, the baseline cannot be used any longer as the reference and guideline for the project course: to try to describe in advance the evolutionary stages characterizing the system is neither effective nor functional to the achievement of the real objective, being ineffective to predetermine precisely the steps required to get to these stages (i.e., create a WBS). An "ideal" snapshot of how the project should be might even turn out to be misleading and, overall, counterproductive. Of course it does not mean that concepts such as objective or baseline become totally meaningless and valueless: they should simply be adapted to the features of the project as a complex system.

As already pointed out, the project is a self-organized system seeking existence at the edge of chaos to create innovation and evolution. Well, evolution toward what? The ultimate goal of any evolutionary process is the achievement of a different and better condition than the present one, which implies the existence of a term in comparison, that is, an objective. For a living being, the objective aimed at by the evolutionary process might be survival, procreation, or satisfying a primary need such as hunger and thirst.

As far as a project is concerned, the objective aimed at by the evolutionary process is certainly its expected result, together with the intrinsic need to build and assert its existence, its "becoming" something which it necessarily expresses, because it is the "creator" of something that did not exist before. The project wants "to be" other than "to do," and it needs to define itself and create its own identity, in order to reach satisfaction.

Embracing the idea of evolutionary conditions, it is necessary to change the mindset concerning the expected results: they are no longer activities to be executed according to planned schedule and cost, but represent a final state to achieve, composed of deliverables and, above all, expectations to satisfy, benefits to implement, and changes to pursue. In this way the end (the result of the pursuit) cannot be confused with the means used to pursue it (the things that must be done).

Identification of the expected results of the project is one of the main tasks of the development of the emerging project charter. As is always the case with this tool, the contribution of all the stakeholders is of the utmost importance, supported by synthesis and convergence, continuously and relentlessly carried out by the project manager.

For a project, the facet of "becoming" perfectly fits inside the emerging project charter, whereas the facet of "being," as previously explained, cannot be found in the baseline. A specific tool is needed, which lets the project entity "imagine itself becoming" and build its own all-accomplished evolution: a project manifesto (see Figure 6.5), both representation and metaphor of the self-consciousness that the project (and the project team) wants to achieve and to aim at in the course of "creation of possible worlds." The project manifesto collects all the expressions that shape whatever the project wants to become in order to create the banner under which the "project system" recognizes itself as unique, dreaming, acting, and disclosing itself to the world: symbols, metaphors, values, and situations.

In principle, no constraints are set on the content of the project manifesto, provided that it identifies and represents the state of "existence" aimed at by the project: the state which, together with the expected results, represents the true objective to pursue, the constant reference that drives any action at any time. It does not mean that the project manifesto cannot be changed, because the self-expression pursued by the project can change during its life cycle, as a consequence of the events and of the conditions that have characterized the evolution course until that moment. This is typical in all the complex systems that co-evolve together with their context.

FIGURE 6.5
Project manifesto: project wannabe.

On the other hand, it is clear that any variation in the project manifesto calls for careful consideration, sharing, and convergence among all the involved parties, so that the whole system knowingly steers toward the goal of a new self-consciousness. It is not a matter of "authorizing" the change of objective, but rather of allowing the new evolutionary needs to emerge in the right direction: these needs have been generated by the evolution of the events, and they must be implemented in order to achieve the final goal successfully.

The project manifesto, together with the emerging project charter, can be used to develop the framework to integrate the identification of both the objective and the result to pursue (where we want to get), the definition of organizational scope (how do we operate), and the self-consciousness of the project (who do we want to be). These three dimensions are tightly interconnected and they all express the essential aim of project building, the engine that embeds the driving and governing capabilities of project driving.

The baseline as a general guideline to follow (how do we get to the final result) loses its value as a reference, because the focus has shifted from the course to the arrival point. It is no longer important to know whether the project is progressing as forecast, but it is important that the status of the project is aligned with its self-consciousness and with the result aimed at, that is, with both its project manifesto and the emerging project charter, which are fully entitled to become the most significant benchmark for the evaluation of project performances.

Project Systemic Map

As discussed in the previous paragraph, when we consider a project as an adaptive complex system the concepts of objective, charter, scope, and baseline should evolve into the ideas of emerging project charter, and project manifesto, tools whose aim is to seize the essence of what the projects want to be rather than simply characterizing contents.

Moreover, the definitions of "inside" and "outside" the system turned out to be meaningless for the identification of the boundaries defining the project. And we learned that the possibility for a project to achieve its objective should not be evaluated with respect to its content. Anyway, it does not mean that it is useless to define project content or to identify its features and its peculiarities, not at all!

As already pointed out several times, even the work breakdown structure, the main content-identification tool used in the traditional approach, clearly shows its limits. "A deliverable oriented hierarchical decomposition of the work to be executed by the project team to accomplish the project objectives and create the required deliverables has limitations. It only organizes and defines the total scope of the project"; this is the WBS, according to the PMBoK. But in a complex system it is useful to represent all the different types of its characterizing relationships, first of all the nonlinear ones (action and reaction) which are so important in the connotation of the complex nature of the system: WBS can only represent hierarchical relationships. Moreover, the same hierarchical nature of the WBS prevents multiple correlations between components (the hierarchical concept requires each element of the structure to be referred to one and only one upper-level element), thus excluding the possibility of seizing and highlighting any form of redundancy, and also the same inclusive nature of the AND that are typical of complex systems.

Indeed, the concept of hierarchy expressed by the WBS is still valuable even for a complex system, inasmuch as it is an expression of a synthesized vision. And this is even more valid in a complex system than in a complicated one. Also the possibility offered by the WBS of considering the project at different levels of synthesis or detail is a nice-to-have asset in a complex system, as well as the possibility of representing relationships (even if limited to the hierarchical ones only), which is the real powerful feature of the WBS.

In order to identify and represent the content of a project in a complex context, the ideal tool should be able, first of all, to highlight each and every component (see Figure 6.6): both tangible and intangible elements,

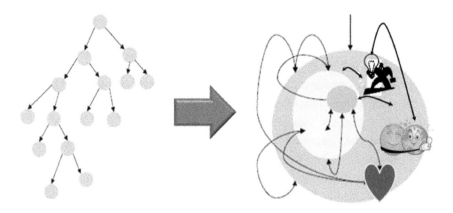

FIGURE 6.6
Project representation: from hierarchical to systemic.

connections of whatever nature, behaviors, impressions, feelings, and anything useful to understand and define a snapshot of the nature of the project.

Such a tool should allow redundancy, should maintain hierarchical logic without a rigid hierarchy, and should be something that *provides a representation of the project allowing a deep and full understanding of the true essence of the project.* It should also be easy and ready to use, as it would be the main tool to manage the relationships among all the stakeholders, project manager included, and the project.

A project systemic map should be developed, a map describing project content where it is possible to freely include whatever is deemed pertinent to the nature and development of the project at a given moment, including things to do (WBS may be part of the project systemic map), moods, rumors, jokes, milestones, budget, useless and outdated information, conflicts, successes, stories, dreams, and anxieties. Mind maps can surely be of great help in the representation of a project systemic map, because they have all the required features and, on top of that, are easy to use and to understand.

Defined as such, the project systemic map can be used to describe the project, to know it more deeply, and to represent its evolution. The point of view of the stakeholders can be shared and unified; diffused leadership is eased up (see Chapter 9). It helps to find one's bearings, to understand, to find out, and to make decisions. It can be used to identify the early and subtle warnings and opportunities, to exercise inclusion (AND), to "see" the intangible, to spot the emergences (i.e., what emerges) more easily. It is a tool to collect and integrate information. It enables systemic vision, while supporting the analysis of details whenever required.

The project systemic map cannot be understood as a stable reference, because it continuously co-evolves along with project execution; its importance is twofold: it is a means of communication between stakeholders and the project, and it is the way the project represents itself. For the project manager it is the tool *par excellence*, a comprehensive map to reliably sail through the project, to determine its state, its evolutions, and its changes. Thanks to it, the project manager can exercise the role of promoter and let the project achieve its results, described by the emerging project charter, while acquiring its self-hood expressed by the project manifesto.

PROJECT DRIVING (GOVERNANCE)

One of the strengths of the project systemic map lies in its ability to embed each traditional management tool, including the WBS, in a thread that can be used to lead the team through worlds of complexity. The project map can be full of connections and be extremely complex, as much as the project itself, exactly because it is a consequence of a project building action, which tends to take everything into account.

In addition to their representation of fleeting accuracy, complex projects need driving mechanisms, a governance, to provide a high-level synthesis of what we are planning and what is going on both in the life and in the world of the project. The reason for the term "project driving" is related to the tools that the project manager can use to drive the project toward the final objective, governing a dynamic situation in which changes are not always under control. In this sense the project manager is a leader rather than a manager of the project, who is able to choose the right way at each bifurcation.

In order to devise the tools appropriately, we have borrowed some concepts from both map theory and class/object theory, together with corresponding state diagrams. By integrating these two worlds, we have merged two separate tools into a single one: mind maps, devised to adapt to the dynamic of the human brain, and class/objects, from information theory [3]. Two separate entities can be identified:

- Emergences
- Domains

Each entity has its own elements, which are interrelated as if they were on a map, and which feature properties and states just like classes and objects [3].

Emergences Diagram

In order to seize the right hints in the project systemic map or to "write" inside it about project evolution, we should enable relevant project topics to emerge, those facts either influencing the evolution of the project or calling for a strategic decision: complexity and dynamism, multiple relationships among people, actions, and plans.

Time plays a fundamental role, because it is usually limited (therefore being a constraint), and because as time goes by, both scope and charter need to be either preserved or changed. The traditional project management approaches are more focused on the verification and preservation of time, whereas the complex approach puts the emphasis on the evolution of the project toward alignment with a final objective. A driving mechanism should take dynamism into account to avoid the "control pitfalls": in other words, it should spot "project emergences" and help the project manager in turning emerging changes and risks into opportunities.

Current best practice requires a continuous alignment of the project with both approved changes and the scope, and it is a way to apply linear feedback to the project, aimed at bringing it back on track, reducing the risk of divergence. This continuous activity is quite expensive and it can potentially hide those context changes that could be either exploited as opportunities or seized as driving needs of the project itself.

Therefore, it is necessary to list and to incorporate into a dynamic relationship several things: hints, specific elements or events, partial results, external facts, and project map: that is, correlate "emergences" with the project systemic map and use them as decision triggers. But what do we exactly mean by emergence? These are facts that "emerge" during the life of the project, and may be important but are not necessarily urgent. It is a way to represent news, and we have tried to identify those peculiar features of the emergences useful for our objectives:

- Emergence is a new fact showing up on a project's horizon, which should be looked for especially, but not exclusively, among the weaker elements that are traditionally poorly tracked inside project documentation.

- An emergence is either positive or negative depending on both the context and the response to the emergence itself. In the traditional project control approach, emergence means steering away from the objective, whereas in a complex context each emergence should be evaluated as positive or negative with respect to both the situation and the ability to exploit it.
- Emergences are not urgencies.

The last feature is of paramount importance in this discussion: urgency is always associated with a "red alarm," a negative event, an action requiring immediate reaction. Emergence implies attention to something that might change the project's horizon, something to take into account. Usually it is just a feeble warning related to human and relational factors that are worth considering and representing in the overall picture to keep them under control. Of course, an emergence might stimulate "urgent action," but it is not always the case.

What is the relationship between the project and its emergent characteristics? If we imagine that the project is a living system spread across a two-dimensional plane, emergence is nothing more than corrugations and protuberances on the project's surface, new surfaces that emerge and are brought to our attention to be evaluated. In this sense emergences can be compared to the biodiversity elements of ecosystems. Biodiversities emerge over time, they can be either permanent or temporary, and project emergences are just the same. Stability over time of this new agent (the emergence) modifies the ecosystem itself (the project) and influences the context. Without persistence, on the other hand, the emergence disappears (the new agent dies out).

Developing the comparison further, it even becomes acceptable to talk of the morphology of the project [4, pp. 2–6], replacing this concept with the traditional one of project structure. Variation of morphology, that is, of the most important characteristics, takes place when emergence becomes a permanent element on the project's horizon.

The whole set of actions changes its characteristics and is redirected to take into account the new scenario to keep on course (here lies the driving ability of the project manager) toward the objective of the project. In order to get to the final objective, full compliance with both the emerging project charter and project manifesto, the two activities are equally important: identify and contextualize all the emergences and correctly plan activities.

In the long run, leading the project, keeping in mind the objective and acting according to emergences, and establishing missions [6, p. 105] for the team (immediately effective and limited in time), might turn out to be less difficult and more truthful than developing a detailed plan. In fact the latter might be valid for a short period in a turbulent context whose time horizon is wide.

Which are the emergences? It is absolutely the responsibility of complexity-aware project managers to identify them first, and then evaluate them together with the team. Many projects have experienced many emergences, both macroscopic and almost negligible: faulty third-party software modules that are used in our project, an opportunity to get a bigger order because a competitor has gone bankrupt, changes inside the project team due to company organizational changes, change of reference contacts inside the customer's company, finding a conflict of interest, the rise of a conflict between team members, an earlier completion of a phase (which sometimes opens up unpredicted scenarios and potential risks, etc.). Every kind of project emergence that any team member might deem as relevant, even if apparently of minor importance, should be "recorded" in a diagram (see Figure 6.7) as if it were an object, together with the event date, its properties, and its relationships with other emergences.

The more the project manager is able to seize and correlate the elements, no matter how small, the more effective the diagram (see Figure 6.7). "Emergence #1: the contract between the customer and the current

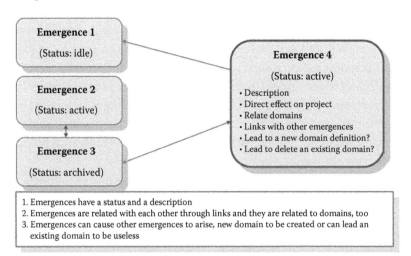

FIGURE 6.7
Typical scheme of an emergences diagram.

supplier will end sooner than expected." "Emergence #2: a strong competition arouse between two team members." "Emergence #3: clash exists between two relevant stakeholders." "Emergence #4: competitiveness has caused some activities to complete earlier." "Emergence #5: a kinship exists between a team member and a customer's stakeholder." "Emergence #6: my team's consultant will be available two weeks earlier."

Typical questions arise. May I exploit an anticipated closure to either start other activities or obtain new orders? Can I generate a positive feedback regarding the rest of the team? In addition to competition, are there other aspects, envy, maybe? Is the project affected by envy? Are increasing internal conflicts due to envy?

The objects in the diagram have corresponding states that highlight their relevance at the time of the observation. If the emergence is somewhat overcome by other events, or if a relationship between two emergences is no longer present, then the emergence is not removed, but is placed into an "idle" state. Dealing with complexity, nothing should be left aside (project narration is an important factor, as well as redundancy, which facilitates the survival of the project), because anything can be useful at the proper time. At the same time, we want to highlight only what is relevant at a given moment, to keep the right focus on decision making.

Once again, it is useful to rely on the comparison with the most analyzed complex systems: the ecosystems [5, pp. 2–6]. Emergences are similar to the agents of a biological system, limited both in organization and in resources, which communicates with the external world: agents continuously evolve both in number and in type, even if available resources (biomasses) are limited. Dynamic descriptions of the agents are similar to the concept of stakeholders' evolution, and also to the idea of emergence. Indeed, stakeholders' evolution can be represented in the emergences diagram.

Summing up, the emergences diagram is an aid for both the team and the project manager to seize, highlight, correlate, and evaluate the impact of what is new inside the project, offering a systemic approach to the active management of unpredicted changes. Though embracing a wider scope, it is a complement to the risk management plan, which focuses on minimizing interference with the project. The diagram also supports project narration: its ability to store the events generates useful information redundancy. The project systemic map and the emergences diagram are valid tools to manage the project dynamic, as well as weak and strong relationships, integrating the traditional view of the project, and adapting the concepts of scope, WBS, and risk to the complex context.

The Space of Domains

When the feeble warnings and information can be synthesized in a dynamic overall picture, the result is more efficient project governance. As human beings, we naturally tend to simplify in order to reduce the number of variables and seize the essential concepts, understand the situation, and manufacture our own experiences. That's why we build a model of reality that is static, simplified, and manageable, and we use it for our forecasts and conjectures. The complex approach states that minor elements should not be neglected, because they are the real indicators of a change, and they can be decisive for the success of the project. Throughout the project life cycle, a huge amount of feeble warnings and hints arises, either from emergences or from the evolution of the plan, and it is a barely possible task to consider and manage them all. Oversimplification is not possible unless we want to fall back on a traditional model; therefore a new synthesis strategy taking into account complex factors is needed. The solution is to avoid static modeling completely, inasmuch as it would never be accurate enough to be valid, but just develop a synthesis of the project which should be manageable, assessable, and modifiable to adapt to the context: this is the key differentiator with respect to the traditional methodology described in the PMBoK.

Although the emergences diagram is useful both to keep track of all information and to highlight relationships between facts and project that cannot be modeled a priori, something else is needed to view and evaluate both emergences and the project systemic map together, with the purpose of making decisions, planning, and executing proper actions and evaluating the results of actions.

The tool to be devised will help us shine beams of light on the project from different perspectives, depending on the current scenario, which dynamically changes. One such spotlight returns an effective and synthesized view of the project. By carefully choosing some specific point of view, the project manager creates domains, each of which represents a comprehensive view of the project that can be used to verify both the health and the evolution of the project with respect to that point of view. The domain includes the evaluation of all the elements and relationships pertaining to the chosen point of view, coming from the project systemic map, the emergences diagram, the emerging project charter, or the project manifesto by means of a proper set of indicators describing a particular point of view on the project.

This is a suitable way to synthesize complexity and to get to the ideal target of understanding the meaning of the myriad variables and relationships embedded in our project. In other words, the motto for this tool could be the following: domain does not mean the exercise of power on the project, but rather an overview of it. The set of domains, each one including properties and evaluation criteria (KPI, described hereafter), all of which interconnected by relationships, is defined in our representation as the space of domains.

A domain, like emergences, is represented by an object that has both properties and states. But emergence is a single fact that has relationships with other facts and relations, whereas the domain is a product of synthesis offering a specific view of the project, an evaluation of effectiveness, and a progress status with respect to that view. Representation technique is identical, a mix of map and state diagram, but its meaning is more general than the emergences diagram.

Which are the most suitable points of view for our scope? When answering this question, freedom, sensitivity, and professionalism of the project manager is put at stake; the project manager has to prove the ability to involve the team in the creation of both the domains and the indicators, and also to be capable of setting the real standards and benchmarks (sometimes different from the ones declared by the stakeholders) against which the project should be evaluated. This tool also gives the opportunity to consider a key concept of a project managed in a complex context: the real measure of the success of the project should be performed not only against the specification parameters, but also against true compliance with the objectives, including those implied in stakeholders' wishes, but never expressed openly.

The relationship between the selection of the proper observation domain and project success would deserve a separate dissertation, which we delegate to others. The space of domains (see Figure 6.8) complemented by the set of evaluation criteria also known as KDI, key domain indicators, provides the project manager with a framework to collect information dynamically.

Similarly to the case of the emergences diagram, the choice of relevant domains not only depends on the project manager, but also on timing factors. In this sense there is a tight correlation with the topic of the identification of the right time to act in the project (propitious time). At a given moment some domains are relevant, but they can be different over

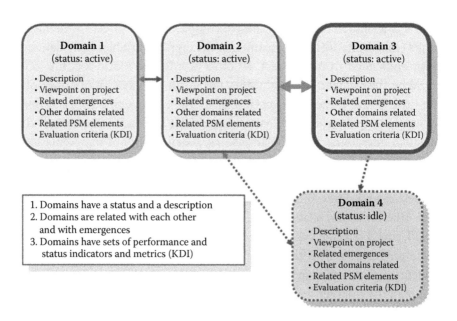

FIGURE 6.8
The concept of space of domains.

time. The function of the space of domains and its corresponding KDI is to focus both the attention and the action of the team on topics that are fundamental to both the stakeholders and the final objective of the project at the time of the analysis; also, another function is to allow a correct evaluation of the health state of the project, thanks to the aggregation of relevant information. The space of domains is both dynamic and symbiotic with the emerging project charter, with the project systemic map, and with emergences, thus the state of the project is constantly adapting the actual objectives and the actual situation (variations in both domains and KDI), and the project history is fully considered and never neglected.

Examples of domain might be: the climate of co-operation (whose KDI include: the number of contrasts, the degree of information sharing, reciprocal help, the reduction of costs due to absenteeism, personal productivity, etc.), report documents (whose KDI include: customer satisfaction, the lack of ambiguity, etc.), customer satisfaction (whose KDI include: the number of escalated issues, the number of new orders, etc.), project effectiveness (whose KDI include: rework and change request rate), and the integration of two products. Even the satisfaction of the customer's sales manager might be a valid domain, even if he is not directly involved in the project.

PUTTING EVERYTHING TOGETHER

The project systemic map, the space of domains and the emergences diagram are tools that can be used individually to manage a project. It is also possible to use them together whenever an integrally complex approach is advisable. There is a leitmotif underneath these tools, which allows their combined utilization (see Figure 6.9).

A beam of light illuminates the vision of the project for the stakeholders: a domain is just another specific beam of light on the project. The space of domains is therefore the totality of spotlights that enlighten the project we have identified at a specific moment, embracing both the emergences and the project map. In this sense, they generate a synthesis that is partial, measurable, and dynamic.

The project systemic map, the space of domains and the emergences diagram can be regarded as three intercommunicating planes, the space of domains being a sort of panoramic viewpoint to observe what happens in

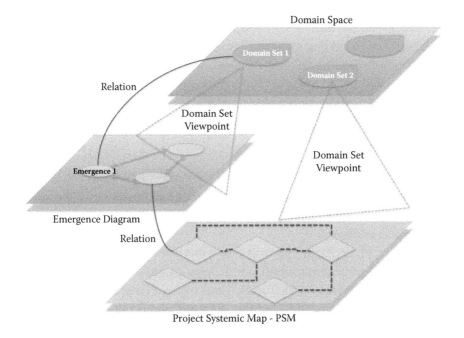

FIGURE 6.9
Relationships among PSM, emergences diagram, and space of domains.

the other two planes, as the three planes are related by threads (the connections). These three tools are dynamic, but are able to keep track of both states and relationships over the project's lifetime: in this way any situation, state, image, fact, or domain can be recalled in order to recover the related experience and results.

Such a rich set of representations can be used to build a consistent and clear project management information system. Guidelines can be derived from both the emerging project charter and the project manifesto; the project can be built out of the project systemic map; changes to the project structure can be managed by the emergences diagram; domains are a way to check the project's growth and to decide how to drive it.

NOTICE TO MARINERS

Sailing across Complexity

Our journey began among experiences of inadequacy with respect to project context; on the way, we found safe harbor in the complexity approach, which allowed us to find ideas and overcome a linear view that was sometimes both misleading and unproductive. We have seen that it is possible to create tools that complement the traditional reference ones, and can be used either standalone or together on a case-by-case basis.

It is the project manager's choice, but some elements can influence this decision:

- The emerging project charter and the project manifesto define the essence of the project in terms of the required final state of "becoming" and "being" beyond the limits posed by the concepts of charter, scope, and baseline.
- The project systemic map together with the emerging project charter and the project manifesto can be used to build the project implementation based on their indications (project building). The project systemic map may include the WBS and should therefore be considered as a widened version of the most renowned project management tool.
- The emergences diagram is used to notate, record, and evaluate the impact of anything happening to the project, in order to understand

any change in scenario, to steer decisions, and to position the events in relation to the project systemic map. In the case of highly dynamic situations, it can even become a governing tool by its own.

- A domain is a beam of light shining on the project, bearing the elements (KDI) to evaluate that particular point of view. Should it be necessary to focus on a specific view of the project, unifying all the elements, parameters, and decisions that have a relationship with it, we can build a domain to correlate that particular aspect with the project as a whole. If a "bird's-eye view" of the project is needed, we can create multiple domains that widen the horizon and create further reciprocal influence relationships with both the project systemic map and the emergences diagram, whenever the three tools are used altogether.

These guidelines are just a suggestion and, of course, are neither exhaustive nor limiting as far as the applicable scenarios are concerned. The PMBoK describes the tools and the processes available to the project manager, who can choose the most suitable ones for a specific project, and then be able to complement such tools with a "complex" approach that can be as pervasive as both the project and the opportunity dictate.

Reducing Complexity

Even if some governing tools are available to help sail through complexity, it would be easier to be able to avoid it. Such a strong statement triggers a question we can ask ourselves: is there a way to reduce complexity in a project? We purposely did not mention a measure of complexity because measuring implies a precise definition of parameters and comparison terms, which in complexity theory are just indicators and signals that something exists. The topics of complexity evaluation and subsequent creation of paradigms of choice for project management tools go beyond the objectives of this book, and they deserve a dedicated investigation.

Complexity features both an objective component and an "apparent" component, the latter related to both experience and knowledge of the context. They cannot be analyzed separately inasmuch as they are closely related, but they can somehow be understood together so that the project manager and the team can arrange a conceptual model and simplify some recurring items, coming both from experience and knowledge. "Turn whatever was unpredictable (due to lack of either knowledge or experience) into something predictable": that is, transforming something "apparently complex" into something which is "really complex" by eliminating "ignorance-induced complexity."

The more an interdisciplinary background is present, the better new contexts and new projects can be faced. It is also true that the more a context is known, the easier its challenges are understood. We have just got to a bifurcation: knowledge and experience are factors that help in reducing complexity, but such knowledge and experience are not always fully available to the project manager or team, because each project reality is different and highly specialized. This is where a co-operative climate inside the team and the ability to introduce an agile leadership [5, pp. 67–68] become helpful. These are indeed factors that help reduce complexity.

We can then imagine the project manager as someone who knows and uses the traditional project language, who is able to carefully choose and wisely blend alternate tools whenever needed, and who above all can facilitate a real climate of diffused leadership, the best and most effective tool to reduce complexity.

Going back to the initial sketch of PM-God, which is a strong picture for the project course devised by Francesco Varanini (see Figure 6.10), we foresee a project manager who is still like a pilgrim wandering between the area of complexity ("casting bread upon the waters") and the area of determinism and forecasting ("build one stone over the other"), but for whom the burden of the Cross is lighter than before, because such "being God" is more often shared with the other pilgrims.

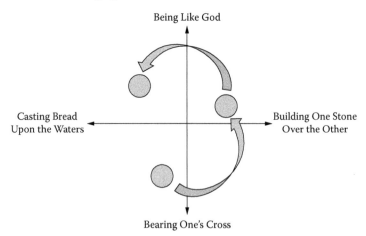

FIGURE 6.10
Pilgrimage between the different states of being a PM.

GIVING IT A GO

The tools described in these sections can be used in several ways, depending on both the context of the specific project and the preferences of the project manager, who can use either one, some, or all of them.

A Couple of Suggestions We Would Like to Offer

First of all, a hint to adopting the tools, in case more than one is chosen: integration, from the logical, informative, and technological points of view, must be guaranteed as much as possible, in order to let the tools express their full potential in a complex context. Integration is the key to that systemic view mentioned before, inasmuch as it is the only way to seize those feeble signs that often determine the emergent success of the whole project. The second suggestion concerns the technology supporting the tools: a huge amount of information needs to be collected and correlated; therefore a proper information management technology is required.

Of course, when we compile the project systemic map and the emergences diagram, we add both objects and relationships (what we see, or we seize, or we imagine of our project is composed of items and their relationships): therefore we predefine the content of the tool. But if we could use an information management tool able to dynamically create new connections between the objects (derived from the attributes of the objects themselves), it would result in the creation of new knowledge, which is a fundamental prerequisite to exercising the governance we discussed before.

Web 2.0 technologies are a perfect example: tags can be defined on the fly to create structures for data analysis depending on the current need, rather than predefined (e.g., pictures can be tagged in order to create multiple photo albums dynamically; e-mail can be grouped by category rather than by folder, and the same mail is referenced multiple times in different contexts without the need of duplicating it).

Reducing predefined structures for archived data organization not only eliminates physical storage limits, because everything can be kept, but also opens up new horizons to the way the knowledge can be built, in addition to enabling successful projects and making the project manager's role more rewarding and happier. The use of concepts such as data mining, ontology, and semantic web is an additional incentive to keep on sailing in the increasingly more intriguing world of complexity.

REFERENCES

1. Project Management Institute (2008). *A Guide to the Project Management Body of Knowledge—PMBoK® Guide*. Newton Square, PA: Author, Chapter 3.
2. De Toni, A.F. and Comello, L. (2005). *Prede o ragni? Uomini e organizzazioni nella ragnatela della complessità*. Turin: UTET.
3. For a detailed description of the state diagram, please refer to: UML http://it.wikipedia.org/wiki/Unified_Modeling_Language
4. Keymer, J.E., Fuentes, M.A., and Marquet, P.A. (2008). *Diversity Emerging: From Competitive Exclusion to Neutral Coexistence in Ecosystems*. SFI Working Papers. Santa Fe, NM: Santa Fe Institute.
5. Giancotti, F. and Shaharabani, Y. (2008). *Leadership agile nella complessità. Organizzazioni, stormi da combattimento [Agile Leadership in Complexity. Organizations, Fight Formations]* Milan: Guerini.

7

Stakeholders' Worlds

Mariù Moresco and Carlo Notari

CONTENTS

Looking into each globe, you see a blue city, the model of a different Fedora. These are the forms the city could have taken if, for one reason or another, it had not become what we see today. In every age someone, looking at Fedora as it was, imagined a way of making it the ideal city, but while he constructed his miniature model, Fedora was already no longer the same as before, and what had until yesterday a possible future became only a toy in a glass globe.

The Invisible Cities by Italo Calvino

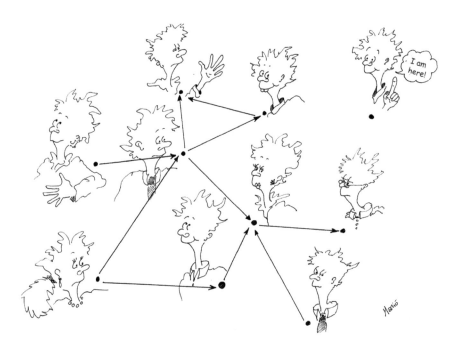

INTRODUCTION

Once the objective has been clarified in one's own mind, thanks to the initial conversations with the sponsor, the next step is to share it with both the customer and the several people involved, aligning their expectations to one's own and vice versa.

This is the moment when the project manager begins the difficult journey toward identification and analysis of the so-called stakeholders, that is, the people, entities, and organizations who have whatever interest in the project. Such an audience can be more or less widespread depending on the project: it usually bears several different interests, quite often contrasting ones. It is worth recalling that a successful program manager not only has to comply with schedule, cost, and quality (as pointed out by every project management handbook), but he or she also has to give the stakeholders a successful perception of the project.

Traditionally, the approach to analysis and management of the stakeholders is limited to mere identification together with some hints on the method of neutralizing potential "enemies" and emphasizing the positive influence of potential "friends." On the other hand, when complexity and uncertainty of scenarios increase, the traditional approach becomes totally inadequate and the project is bound to fail due to a series of hurdles and difficulties unless a more comprehensive and systemic method is followed.

But what does complexity mean in this context? Obviously, speaking of people, entities, and organizations, management means relationship skills, the way of interacting with each other, understanding individual facets, reference culture, and the relationship network's context, with the aim of getting the best out of the reciprocal points of view in order to blend them and make the project a success for everybody and not just for someone.

For example, if our project deals with the installation of a transmission plan for a mobile network, it is clear that the complexity of the different types of stakeholders is really significant. In fact, we should manage:

- The land owner (whose unique interest is profit maximization)
- The organization in charge of the project (interested in the maximum efficiency and effectiveness of quality of service)
- The entities in charge of providing authorizations (interested in control, but not interested in either cost or quality of service)
- Consumer organizations (interested in health, quality of life, and of service, scarce interest in cost, high interest in process)
- And so on, for each stakeholder involved

Therefore the project manager must be technically capable of interfacing with stakeholders correctly (using skills available through the technical team), but above all be able to build relationships.

Starting from these assumptions, which are the tools and approaches that a project manager needs to handle relationships as satisfactorily as possible throughout the project lifetime? First of all, the context the project develops in must be carefully analyzed, that is:

- Contact points between PM, the organization and the project
- Strength and weakness points in the development of the project
- Potentially successful and unsuccessful strategies

It is important to remember that the project manager is not only a guide able to select tools and techniques, but also, and above all, a reference point who co-ordinates performance expectations of every member of the project. In fact, the so-called performing organization asks the program manager to be effective in efficiency, which is gained by means of involving, communicating, listening, motivating, organizing, and coaching. All the above-mentioned items call for the smart and correct management of every stakeholder of the project.

STAKEHOLDER AND PMBoK

Before investigating the above-mentioned approaches further, it is important to summarize the way the PMBoK® (one of the main reference books for project management [1, pp. 23, 246]) deals with the topics and the issues related to stakeholders, which occurs in two separate sections:

- In Chapter 2, which deals with "Project Life Cycle and Project Organization," where an entire section is devoted to the explanation of who the stakeholders are, how they can be divided into macro groups, developing some of the most important examples (customers/users, sponsor, project portfolio managers, project management office, program and project manager).
- In Chapter 10, which deals with the communication knowledge area. In particular, in Section 10.1, the process called "Identify Stakeholders," depicted in Figure 7.1, is explained.

The description of this process recommends identifying these stakeholders as soon as possible, to discover their potential influence on the project, whether it is positive or negative, to understand each individual degree of interest in the project, their expectations, and so on. Among the input

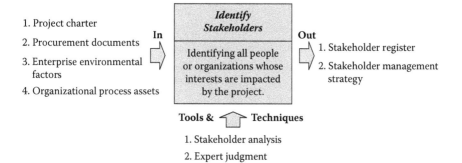

FIGURE 7.1
Identify stakeholders according to PMBoK.

of the process, the following can be highlighted: the project charter (i.e., the document providing information on what has to be developed), documentation about supply contracts (suppliers are important stakeholders), company culture (which can greatly influence every relationship), and last but not least, procedures (defined as assets of organizational processes) that dictate the rules for managing relationships inside and outside the organization. Among the output and tools and techniques, the following items are mentioned: the elements representing information, the way of collecting them, and the factors to be taken into account for stakeholder analysis.

Paragraph 10.4 inside the same Chapter 10 describes the process of stakeholder management as depicted in Figure 7.2. Even in this case, the manner of representing information and the specific skills required to better understand stakeholders' expectations are described. But how can these skills be developed, and how can such information be retrieved? This

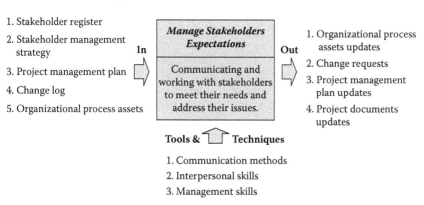

FIGURE 7.2
Manage stakeholders' expectations according to PMBoK.

is where complexity creeps into the project and, specifically, the complexity of this kind of management. In fact, when dealing with "stakeholders' worlds" and assuming the initial question is to what extent does complexity show us the limits of the traditional approach, we should ask ourselves which is the emerging topic of complexity.

An initial indication could simply have been, "Identify stakeholders and collect their requirements," but today this is no longer enough: in fact, after identifying them, it is necessary to know them thoroughly to be able to get into a deep relationship with them, thereby establishing effective communication. In other words, the project shall be considered as a real temporary social organization characterized by interactions among interests, knowledge, and cultures of all the stakeholders involved.

The scope of a project should be therefore considered as the result of all the possible overlaps [2, p. 104] between different "worlds" pertaining to each different stakeholder (see Figure 7.3). Furthermore, it is important to consider that the analysis of stakeholders' worlds cannot be limited to an initial snapshot, no matter how realistic, but it should monitor evolution over time. Complexity of the organizations indeed calls for an analysis that takes into account relationship dynamics (both in individual and in group perspectives), cultural framework, and individual creativity.

The understanding of cultural dynamics helps build the tools to interpret reality and to share knowledge which are also useful in preparing the individual for new and more complex challenges. But how? And which methods and tools should be used? A different stakeholder map should

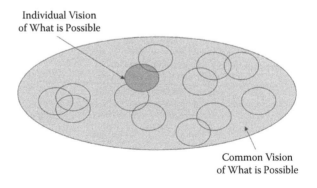

FIGURE 7.3
Individual and group perception.

be devised, which is not limited to the evaluation of a specific influence and interest, but which explores new dimensions: history, practice derived from experience, cultural context, and organizational context, especially the informal one, within which they move and work. And the evolution of all these aspects should also be monitored over time.

Both the evolution and the outcome of a project are strongly influenced by the iterations inside the social group involved, its culture, and the individuals that compose it: each of these three entities, tightly and dynamically correlated, is a phenomenon with its specific features. In this context, and in this chapter, we mainly deal with three huge driving aspects: ethnography, informal networks, and type-watching.

The first two topics show which subject should be considered when dealing with interproject communication, the third one offers an approach that helps seizing the real attitudes or tendency of the individuals (and, therefore, of the organizations they belong to) to develop and optimize, indeed individually, both interchange and relationship.

ETHNOGRAPHIC APPROACH

Studying Organizations as Cultures

The aim of the ethnographic approach is to recognize and analyze the cultural codes that are typical of a group together with its context, in order to understand the reasons why certain phenomena occur. It also allows us to study the logic governing the group being observed; such logic might be different from our own, but it is nonetheless valid.

The term "ethnography" comes from the Greek word *ethnos* which means race, group of people, or cultural group. The union of the words *ethnos* and *graphos* generates the word ethnography, which is a subdiscipline of anthropology aimed at describing the way of living of an observed group, interpreting the meaning of human behavior. The ethnographic approach leads to a better understanding of the culture of the social group we operate in, and allows us to highlight both its dynamics and interactions. The ethnographer joins the team to be studied and uses several research techniques, including observation and interviews, in order to acquire information useful both to understand the group's culture and to explain the observed phenomena.

The sources for data collection can be grouped into three different categories: traces of organizational life, collective events, and the subjects themselves, both in their "natural" behavior and in the one triggered by the researcher. The tools for data collection available to the researcher are the analysis of written text contents, observation, and ethnographical interview [3, pp. 104–118].

According to the Polish anthropologist Bronislaw Malinowski, the ethnographer should "observe while participating," that is, establish an empathic relationship with the subjects in order to grasp their point of view and therefore their vision of their own world. To think anthropologically and to perform field research imply considering every facet of social life of the culture under study, in order to uncover the various meanings that the observed phenomenon entails when it is observed from different points of view.

It also means to recognize and to contrast the fact that individuals sometimes believe that their own behavior and values are better than those of others, or at least can be considered as a safe harbor. If this kind of approach is chosen, the concept of culture must be further investigated. Culture, in its wide ethnographical sense, is the complex set that includes knowledge, beliefs, art, morals, laws, habits, and any other skill and habit acquired by man as a member of society.

Paraphrasing Aristotle, according to whom "man is a social animal by his own nature," we can also say that "man is a cultural animal by his own nature," inasmuch as he fully develops his potential because he lives a cultural experience, and because most of human behavioral models are either acquired or learned throughout the lifespan of an individual. Man is mainly a self-made being, and not a raw product of nature, because the cultural forms of his behavior greatly differ from natural and instinctive ones. As a member of a group, man is different from an isolated individual. As a member of a society, his degree of freedom decreases: a man within a group can be led to believe the most incredible things, can carry out cruel and abominable actions, and can undergo the wildest prohibitions in order to adhere to his traditions and rites.

Even the simplest people living on remote islands are characterized by extremely complex organizational forms and habits. People who live in huts and get their food by primitive hunting and fishing techniques have been induced by fear and ambition to devise elaborate forms of art, myth, and rite: they have tried to keep both the environment and their peers under control using any sort of expedient in order to dominate the world around them.

It is worth noting that every culture is usually more complex than would be necessary to ensure the survival of society. For instance, native Australians have developed their social organization to an extremely sophisticated level, while paying little or no attention to the importance of technology. This is related to the relative value that a society assigns to different topics: which are the activities that a society really considers important? Why? Is it possible that some activities give a small practical contribution, but at the same time match the psychological needs of the individuals, for instance, the need for reputation and respect by others?

Activities that are valued most will generate better rewards in case of success, and whenever new potentials are achieved because of them. Quite often, elements of superstition can even be found in an advanced cultural context. From an anthropological point of view, superstition is a mechanism of both defense and reassurance used by both individuals and groups to justify their failures and their uncertainties, which are of course more relevant in a period of uneasiness and uncertainty.

We could even say that superstition is a necessary evil, inasmuch as it is useful in case of existential crisis. It is also interesting to note that, from an historical perspective for either an individual or a group, superstition sometimes represents a boundary between a past and a new era, a clear cut with the past in order to better define a new identity.

Both dynamism and change are embedded in human culture, even if they occur in a framework of cultural stability. Sometimes, changes seem to be vast and dramatic for the members of the society in which they occur, but quite often just a small part of the cultural context is affected. Change can be generated either inside or outside a group: it develops internally in the case of either a discovery or an invention, and it comes from outside as a result of a process of "adoption," that is, of cultural contamination, which is never indiscriminate, but rather selective. New cultural models are accepted as long as they are in tune with the old; in other cases, they can even be modified in order to match existing cultural context.

When a form of culture has been created and positively applied by the members of a society, it is also imitated by the new generations, both directly without any instruction and by means of verbal instruction and conditioning. Even if culture is the instrument by which human beings adapt to their own context, it does not necessarily lead the individual to a passive behavior: the individual keeps his capability of thinking and of devising new forms of behavior when facing new situations for which existing cultural models are inadequate.

Adaptation is indeed a circular and never-ending process: it is an inter-action between the individual and his group which is favored by creativity, which is both the fundamental expression of an individual's unrest toward the stereotypes of the group and the way to exercise his expression without breaking the fundamental rules of his culture. The individual represents a key variable for any social group, and any innovation element of a culture can be traced to the mind of some individual. Every culture is fundamentally conservative, because it is the set of knowledge developed by a group, above all with respect to a common way of action, rites, and myths that proved to be effective in the past for the survival of the group itself.

If we want to act as promoters of the change we must recognize and keep in mind this tendency that the organization has to return to known schemes (which reduces both anxiety and the learning curve), and on the other hand we must remember that within organizations the physiological and cultural dynamics are the real levers for either preservation or innovation.

Ethnography and Stakeholders' Management

The traditional approach used so far has aimed at objectivity of measured data: today, new points of view and interpretation tools allow us to have a vision of reality that is more organic and less taken for granted. Cultural analysis of organizations reveals itself as a really effective tool in those situations where the system of principles and values that influences stakeholders' behavior represents an obstacle or a resource to be strengthened. Therefore stakeholders' maps inside the project should be revised adding other dimensions that take somewhat into account the culture of each individual.

A possible solution is to identify sensitive "cultural parameters" that can give direction to the relationship with each stakeholder. Let's assume to choose, according to the needs and the values to be pursued in actual reality, the following ones:

- Ethics
- Promotion of creativity
- Innovative attitude
- Organizational welfare

Then let's give a quantitative value ranging from "unimportant" (zero) to "very important" (100), passing through a "somewhat important" (50),

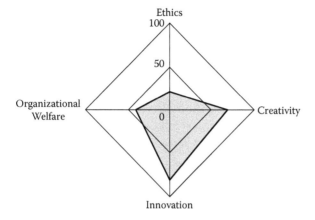

FIGURE 7.4
Evaluation of cultural parameters.

representing it on a Kiviat diagram as shown in Figure 7.4. This graph represents the reference scheme of a "project X" where "Stkh1" is one of the stakeholders. It is easy to see that "Stkh1" belongs to a culture where creativity is encouraged and innovation attitude is very high, but which barely cares about ethical issues and where organizational welfare is not even sufficient.

Such information, which is really useful, leads to conclusions that can be made more realistic completing the analysis with the tool shown in the following section, which can be used to better understand the root cause for the two negative data.

INFORMAL NETWORKS

Features of the Approach

In every gathering of people, the need for rules arises: therefore formal structures are set up inside organizations that emphasize three main topics:

- Hierarchies
- Definition of roles
- Operational mechanisms

The fact that parallel, spontaneous, informal networks are originated is not always taken into account; these are flexible adaptive structures that self-organize internally in order to react to external changes, sometimes with rules which are in contrast with formal rules, but are indeed the real engine of the organization.

It is not uncommon that such informal networks achieve the desired result using strategies that are different from the planned ones. It has been calculated that on average 80% of activities carried out inside an organization follow informal procedures. Understanding the network of informal relationships is therefore important to understand the invisible mechanisms that are often the basis for good practices, defining in detail the tactic required to pursue planned strategies.

The limits of formal organization usually emerge when the context changes and it is no longer the same as the one where formal organization was born and had success. Due to its intrinsic features, formal organization is not able to cope with change efficiently: it cannot handle unstable and unpredictable situations, because its stiffness does not allow ideas, fads, and unexpected behaviors inside its rules.

However, flexibility and adaptation are intrinsic features of an informal network; owing to its informality, a relationship network automatically adapts and mimics actual organizational contexts. Adaptation capabilities of the organizations are always based upon informal organizations that adapt real-time according to the external environment. Formal organization is necessary in order to define roles, tasks, and responsibilities, and it is useful to forecast what needs to be done routinely and how; but an informal network is reactive, and it offers both flexibility and adaptability. It is therefore evident that these network configurations, which are so different, show their effectiveness in different contexts:

- In simple and stable environments, which offer routine services, "rigid" networks work well and are characterized by a low density of relationships.
- In complex and dynamic environments, which offer personalized services, flexible and decentralized networks are needed, with a high density of relationships and a high number of interconnections.

When either innovation or the capability of providing nonstandard performances is a key factor, it is necessary to implement informal configurations of the organization, which are more agile in answering to the increase of complexity and to the unpredictability of the context.

In the case of stable and predictable situations, or in the case of situations where we want to drive toward these characteristics, it is more efficient to have a centralized organization where relationships are not so dense: in fact in this case relationship density can become redundancy, and therefore disturb. So how can we define an informal network? It is definitely a complex adaptive system, a system where a set of actors interacts between them in a way that is not completely predictable.

Representation of Informal Network

This kind of network cannot be represented by means of organizational charts, but rather by means of "connection maps" that represent the learning system of the organization: who knows what, who knows whom, and who works with whom. The informal network is described by the set of real relationships that exist between the actors of the organization and by the specific personal characteristics of each actor.

The basic rule to represent an informal network is to consider only the relationships that are really active. An informal network (see Figure 7.5) can be graphically represented by a set of nodes and arrows. Nodes represent individual actors, and their meaning can be enriched with further information on the characteristic of the individual (e.g., function, seniority, specific skills, etc.), diversifying color and size. Arrows represent the existence of a relationship between the actors and they also show its direction; the thickness of the arrow shows the intensity of the relationship.

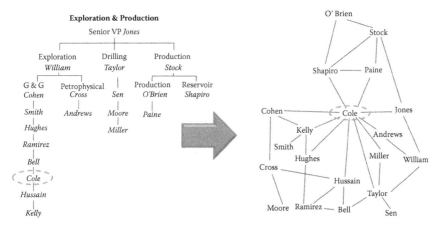

FIGURE 7.5
Example of informal network.

It is worth noting that the perception of an informal network is not as easy as it might seem, even for its actors. In fact individuals often have a distorted perception of their context and in particular of the informal organization, considering themselves at the center of a greater number of relationships with respect to the others. Even the members of the system do not really know the structure to which they belong. Suffice it to think that a social system of 100 individuals can have as many as 4,950 different relationships!

Reading the Informal Network

At this point, it is important to understand how to read an informal network, conscious of the fact that it is not an easy task. The analysis of the network should indeed take into account:

The features of the relationship
The features of the network
The features of the individuals as actors of the network
The features of the individuals as individuals

Features of the Relationship

It is important to note that among the members of an organization different types of relationship coexist and for each relationship a different network exists. The main types of relationships identified within organizations are:

- *Work network:* The relationship of information exchange: who goes to whom to get information on specific topics
- *Advice network:* The relationship for troubleshooting, that is, whom to ask to get solutions (e.g., a skilled technician, who does not share the knowledge) or to understand the framework of the problem (e.g., a coach)
- *Trust network:* The relationship of knowledge of the knowledge, that is, who knows those who have the right knowledge fundamental to create resource exchange networks and to support innovative processes

The first thing to do is therefore to decide which relationship to study, because only one at a time can be considered.

Features of the Network

Once the type of relationship to analyze has been decided, it is necessary to see which features of the informal network grow from that relationship. In this context, the following items should be considered:

- *Density:* Indicates the degree of cohesion of the organization.
- *Diameter:* Indicates the maximum number of steps required to transfer information within the organization.
- *Centralization:* Expresses the degree to which each person is involved to get information.
- *Subgroups:* Identifies informal groups that are cohesive internally, but relatively separated from the rest of the organization.

Features of the Individuals as Actors of the Network

Within the informal network, the individuals inevitably take roles determined by their position inside the context of relationships. It is maybe the most interesting side of the study of informal networks, inasmuch as it leads to the identification of the following characters:

- *Central hub:* Is a sort of concentrator and switching hub who owns information fundamental for that network. It can also become a bottleneck if everybody has to rely on him or her.
- *Opinion leader:* It is an individual with personal charisma and trusted upon by the other actors of that network ("To whom would you go for advice?"). It is very important to identify him or her when dealing with change and innovation processes.
- *Broker:* Acts as a mediator between other people and groups, enabling or controlling information flow, who can promote innovation processes.
- *Boundary spanner:* Tends to stay at the boundary of the organization, linking it to the external world and bringing in new contacts and knowledge. A key figure, both for innovation, and to reduce self-reference inside the organization.
- *Pulse taker:* Has many direct links and is independent of mediators, especially from the brokers, escaping from control and moving in the network as a sort of free agent.

- *Peripheral people:* Individuals who live at the boundary of the network with few significant connections with the rest of the organization. They are often elderly people who could not keep up the pace, maintaining obsolete knowledge, and who are occasional part-time resources.

Features of the Individuals as Individuals

These are the attributes of the actors, which correspond to their specific features, including age, seniority, function, personal skills, professional skills, and so on. The knowledge of the attributes of the actors can be very useful in explaining both position and role within an informal network.

Figure 7.6 shows the main features, and can be used to summarize the various topics involved in the analysis of informal networks. In order to draw an informal network with its real features (i.e., including the analyzed relationship and its actors) a proper methodology should be used, gathering relevant information, elaborating it, and finally graphically representing the network itself. This method, called organizational network analysis [4, pp. 36–61], not only performs the analysis of the informal network, making it visible and interpreting it, but also includes activities aimed at modifying it.

It should be noted that the analysis phase might also involve dealing with sensitive and somewhat confidential data, whose collection shall be performed in a proper way according to appropriate confidentiality policies: for instance, if the task is to understand how a given topic is developed

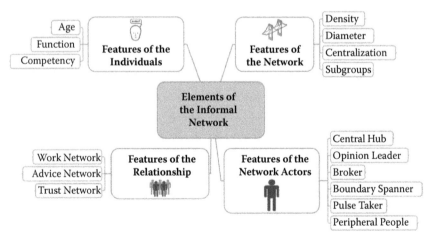

FIGURE 7.6
Map of the elements of an informal network.

inside the organization, knowing the names of the people involved could be avoided; it could be sufficient to get nicknames with no possibility of identifying the people behind them.

Secondly, the attempt of measuring and computing by means of methods and equations an entity that is fluid and all-changing by nature underlies a sort of paradox: as soon as we have accurately measured it, it will probably already be different; therefore the picture will already be outdated.

It is self-evident that such approaches cannot be easily applied to the context of project management in their entirety, but they can certainly be useful to integrate stakeholders' analyses, helping us to position them, to understand, and to simplify their relational dynamics.

Informal Networks and Stakeholders' Management

Similarly to what has been done dealing with ethnography and cultural aspects of the organization, some fundamental parameters can be defined, which can give useful indications about the type of relationship network of each stakeholder. In this way the analysis can be enriched with a more detailed representation that takes the informal network into account. Let's assume that the following parameters are considered in order to provide a measurement of the "agility" of the organization:

- Informality of internal relationships
- Informality of external relationships
- Flexibility
- Attitude to problem solving

Let's than assume that we use a representation and a value range similar to the one used for ethnography: from "unimportant" (zero) to "very important" (100), passing through a "somewhat important" (50). In this way a graph can be drawn, as shown in Figure 7.7. By analyzing the graph, it could be discovered that the poor organizational welfare emerged from previous analysis is probably due to a formal setting of relationships (both internal and external), to a modest attitude to flexibility, even if the ability in problem solving is fairly satisfactory.

By merging the results of ethnographic and informal network analyses, and adding our evaluation about the interest and power of the stakeholder, Stkh1, the synthesis shown in Figure 7.8 can be found. We could then conclude that the organization's ability to innovate is due to the skills and

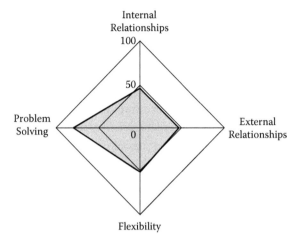

FIGURE 7.7
Example of evaluation of cultural parameters.

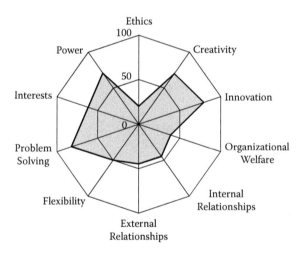

FIGURE 7.8
Example of a synthesis of an ethnographic analysis.

abilities of the individuals, and both internal and external relationships are strongly formalized. Therefore it is important to rely on the individuals, and the tool described in the next section can be used efficiently. It is also clear that the higher the interest and the power of stakeholder Stkh1 over our project, the more accurate our analysis should be.

TYPE-WATCHING

A Short History

Type-watching essentially deals with interdependence and with the way of approaching the "others" in order to get a mutually satisfactory relationship. Through type-watching, it is possible to capture both complexity and diversity using just four dimensions of human behavior. This leads to a constructive answer to the unavoidable classification of the behaviors, and leads to a virtuous development of self-knowledge. Therefore, we talk of observation of human types, rather than classification.

Type-watching can be characterized as a psychological system that is free from judgment and aimed at explaining normal psychology rather than abnormal; it does not refer to "good" and "bad" types, but only to differences, honoring them and using them in a creative and constructive way rather than as a conflict. It originates from Carl Gustav Jung, who suggested that human behaviors are not random, but can be predicted and identified. Jung had a different approach from many of his colleagues because he proposed categories that are not based on psychological diseases and abnormality, but are the result of behavioral preferences related to basic functions that each personality develops throughout life. Therefore every issue that arises during the life of everyone translates, according to Jung, into precise features of the personality which determine both the attractions and repulsions to and from people, things, and events [5, p. 417].

Jung's theory was revamped at a later stage (during the Forties) by two women: Katharine Briggs and her daughter Isabel Briggs Myers, who developed a psychological tool to explain behavioral differences in strict scientific terms and according to Jung's theory of personality preferences. Therefore they developed the so-called Myers Briggs Type Indicator (MBTI), which has contributed a lot to the spread of Jung's theory.

Type-Watching and Stakeholders' Management

Behavioral development of individuals, as we said, depends on events that occur during one's lifetime, on the context we act in, and on the people we interface with, either on the job or in private life. Let's think about the management of a project, which leads to relating with team members, with the customer, with the sponsor, and with other stakeholders, whose role can be very influential regarding the project's success. We could also find ourselves in a context far from our relational habits, even far from our usual place of relationship. All these facts modify our behavior; after all, we all have precise but flexible features that adapt to the context and situations, at least thanks to our survival instinct.

A building is a good example to explain the development of our behavior during our lifetime: its foundations do not change over time, but the exterior parts undergo both superficial (such as painting) and fundamental (such as renovation) changes. The same is true for our behavior: a person can be fundamentally an extrovert, and it will be an immutable feature over the course of her life, but such extroversion will emerge differently depending on the context and the events that occur; indeed, in some situations and contexts such extroversion might even be disguised, appearing as introversion. The types of behavior considered by type-watching are depicted in Figure 7.9.

The aim of the picture is to highlight the flexibility of our personality preferences: therefore the metaphor of an audio amplifier has been used, where each cursor is relevant to a single tone (different features of the personality) and allows a sound (i.e., one's own human type) to be modulated by means of the four cursors. In the same way, our personality adapts to the context, to the people we relate to, to our specific mood of the moment, and so on. In the following, we briefly show macro characteristics of each pair of human types: for further information on the topic, see the book by Kroeger and Thuesen [6, pp. 26–48].

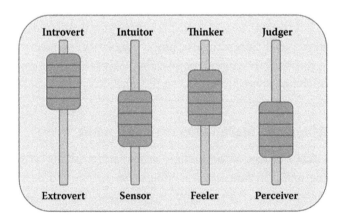

FIGURE 7.9
The types of behavior considered by type-watching.

Extrovert and introvert are the sources of interior energy. The extrovert gets energized and revitalized by the external world, and the activities he or she finds more exciting and stimulating are the same that tend to exhaust and drain the introvert. And the reverse is valid as well: thinking over, introspection, and loneliness that energize, focus, and activate the introvert, tend to exhaust the extrovert. At school, at home, or at work, it is important that both types of individuals can access their own sources of energy.

Intuitor and sensor are modes of information gathering. In this case the focus is on the information gathering function, that is, the endless process used to acquire information from the external world. The sensor is realistic and practical, gathering information from the outside world using all five senses, because it is the only way to grasp every facet of things. Everything is based on facts and associated details. On the other hand, those who prefer to rely on holistic visions, to a higher perception of the things and of their relationships, prefers intuition and is described as an intuitor.

Thinker and feeler are decision-maker methods and relate to the process of making decisions based on collected data. The aim of decision making is to make judgments based on available information, and make decisions accordingly. It is just like eating a good-looking cake: its consistency, softness, and perceived sweetness (judgment) are the elements to decide whether the cake tastes good or bad. In this case, we talk about the thinker, who tends to clarity and objective truth. A thinker considers principles more important than people, so that he often embraces an authoritarian ethic. On the other side we can find the feeler, who is more interested in a person's well-being, harmony, and accord with others. A feeler promotes a humanistic ethic, where people come first and can never be submitted to any end that is posed above them.

Judger and perceiver concern the modality of a relationship. The last area to be considered is perhaps the main source of interpersonal conflicts. In fact it deals with the ways a person usually employs to achieve a relationship with the external world, both from a verbal and a behavioral point of view: that is, one's life orientation, or lifestyle. Also in this case there are two behavioral preferences to consider: the judger is determined and stern, oriented to the goal and to priority assignment. He or she sticks to the project, to the word given, is reliable, and carries out the tasks. On the contrary, the perceiver is undecided but flexible, an improviser, willing to explore, to acquire new information, deferring decisions, and oriented to discovery. A perceiver therefore tends to have several objectives in parallel and gets to the result by refinement steps.

Applying Type-Watching to the Project Team

It is a known fact that most of the success of a project depends on team management, and the team is also one of the main stakeholders. Therefore it is important to facilitate relationships, to understand the inclinations of each component in order to make an effective use of them in the project, and to mediate conflicts properly, promoting both individual and group motivation, and the like. Then it is mandatory not only to understand the inclination of every member, the reference culture, the context of the reference informal network, and how all these things are blended among the team, but also to dig into the points of strength and weakness to harmonize the behaviors.

Effective leadership requires the adoption of a style appropriate to the phase in which the group finds itself. A diagnosis skill is needed (situations, individuals, inclinations, etc.) together with flexible behavior. Effective leadership means to be able to handle both the process that takes the group from depending on the leader to become interdependent, and the process that leads from external to internal control. Because single individualities emerge within the context of the team, type-watching can be the tool that helps the project manager to have a clear idea about the inclination of each individual (leader included) inside the group, harmonizing them toward the objective.

In order to get proper team development, it is important to remember that by sharing control of the group, the leader is no longer making decisions for the group, but rather participates in the decisions themselves: the role of decision maker blends with the role of human resource developer. Another support of positive and sustainable team development is the ethnographic approach, through which understanding of the group culture can be improved, and its dynamic and interactions can be understood. Understanding cultural dynamics is a way to build tools to interpret reality and to transmit knowledge, and it is also useful to set the individual toward new and more complex situations. On top of that, for the project manager it is surely useful to improve knowledge of the theory of social and informal networks, in order to improve the awareness of the human context in which the project develops.

In order to enable an effective development of the activity, the project manager not only has to understand the informal network inside which the project exists, but also has to make sure the team organization is inspired by the principles of the informal network, rather than by hierarchy principles when task and responsibility assignments are considered.

Sharing the Objectives

The project team moves along two important tracks: objectives and action. The first issue the project manager has to cope with is therefore related to the so-called goal setting, that is, sharing project objectives, analyzing, both in qualitative and quantitative terms, each person's contribution. The activities related to this phase of the project are really tricky. In fact, we should remember that the primary cause of project failure is the wrong interpretation of its goals or goal-related topics: understanding expectations and needs, wrong management of the relationship with the most important stakeholders, and so on. This phase should therefore be carried out taking into account that the organization sets criteria for evaluation and measurement of the performance through the evaluation of the degree of achievement of the objectives. The techniques mentioned above represent valuable support of the finalization of the activities of goal sharing.

Talking of type-watching, if for instance the person who has to communicate the goal is an extrovert–perceiver, we should expect that the goal may change or be better defined as his or her thinking (of course verbal, and perceived by everybody) is refined, or elaborates the feedback coming from the others. On the other hand, for instance, a sensor–thinker, who needs to interpret the goal, needs to anchor it to real facts and logic thoughts. It is evident that the project manager cannot use "two-cents" psychology, but has to work first of all on himself to refine the sensitivity and the understanding of his reality and of whoever is around him, in order to get into a relationship properly depending on the personality in front of him.

A successful project manager is able to relate radically differently with different people. That means that on top of the type-watching, she has to interpret the situations depending on the cultural framework and the social network: it is easy to spot the differences, for the same human type, if we consider creative contexts, such as architecture, or other contexts that are highly rigid and coded, such as bureaucratic ones.

THE EMERGING PROJECT CHARTER

Every project manager knows that the project continuously changes: the context changes, conditions change, people often change, and therefore the attitude or approach of some of the stakeholders. This consideration leads to the definition of a tool that is different from the classic ones used thus far: the emerging project charter, a roadmap that collects information which can be used by whoever might inherit our project, such as other colleagues who could have to deal with similar projects in the future. It is important to remember that the project manager should always consider the engagement as a temporary one even within the context of the specific project, and therefore he must facilitate his successor, if any, and also whoever could glean useful hints from the project documentation.

It does not mean altering the concept of the project charter described in the PMBoK, but rather to define an evolution that goes along with it and leads it to an equally important role of reference and guide to be used along the full project life cycle. The emerging project charter is not only a defined project sheet defined, realized, and consolidated in the initial stage, but it also becomes a document that follows all the evolutions of the project while its phases unfold and associated deliverables are created. It can therefore represent a synthesis tool through which the identification of the goal to pursue (where we want to get) is integrated with the definition of organizational scope (see Chapter 6 of this book).

Of course, the ideal solution would be to collect the vision of every stakeholder and put it in the emerging project charter, in order to blend the different perspectives coming from all the subjects involved. But we know that this approach is not only quite difficult to implement (it is difficult to get the description of his or her own perspective from everyone), but it would be even more difficult to co-ordinate, because radically different points of view should probably be blended together.

The solution is to be proactive, describing objectives, significant events, deliverables, and the like (also using the progress report documents of the project), taking into account individual perspectives, sharing our document with each subject involved who has to subscribe to it. Of course this approach is not an easy one. But it would let us lead the game, letting all our skills emerge, together with our knowledge of the contexts (cultures, features, flexibility, attitudes, etc.) that we derive from the use of the tools described in the previous sections.

The emerging project charter should necessarily offer different points of view: information that is significant from one's own performing organization is of little or no importance to the customer; some information is sensitive and cannot be disclosed outside the organization, and so on. It is also evident that trying to implement a valid template for every project is just meaningless and misleading: each project has its own peculiarity and its history, and it develops within a specific context, and therefore the information contained in an emerging project charter is different from one project to another. Basic requirements of an emerging project charter should be taken into account, and developed as the project evolves.

CONCLUSIONS

Summing up, it is clear that complex projects call for a deeper analysis of the stakeholders. We have tried to enhance the so-called "PM toolbox," defining some tools that come from other sciences, such as anthropology, sociology, and psychology. And this is the proof that there is no approach, in any field, that can do without the help of knowledge areas which are apparently far from the topics of the everyday job. Indeed, it is a multidisciplinary approach that creates the so-called "virtuous circles" that lead to innovation.

Kiviat diagrams, previously defined applying basic concepts of ethnography, informal networks, and type-watching, are by no means standalone tools that, once implemented, are meant to stand still for the duration of the project. On the contrary, the evolution of contexts, cultures, historical moment, and people, should force us to keep our analyses updated by means of a very important tool for project interpretation that we have defined as the emerging project charter, which embeds a synthetic, dynamic, and shared vision of the objective at which the project is aiming.

The emerging project charter should reflect the style of its originator, but it should also grant the objectivity of information; it should express the true spirit of the team that led to the generation of the final result. Indeed, it is the tool that little by little ensures the emergence of a shared vision of the project, starting from the individual points of view of the stakeholders.

It is also important to remember that the most important stakeholder is the project manager who is the first one who steers the project by means of his or her cultural context, relationship network, and personality. Therefore behind a successful project there is always a project manager who is fully aware of the importance of self-knowledge.

REFERENCES

1. Project Management Institute (2008). *A Guide to the Project Management Body of Knowledge—PMBoK® Guide*. Newton Square, PA: Author, pp. 23, 246.
2. Battram, A. (2009). *Navigating Complexity: The Essential Guide to Complexity Theory in Business and Management*. London: The Industrial Society, p. 104.
3. Piccardo E. and Benozzo, A. (1996). *Etnografia organizzativa*. Milan: Raffaello Cortina Editore, pp. 104–118.
4. Oriani, G. (2008). *La forza delle Reti di Relazioni Informali nelle Organizzazioni. L'Organizational Network Analysis*. Milan: Franco Angeli, pp. 36–61.
5. Jung, C.G. (1971). *Psychological Types*. Princeton, NJ: Princeton University Press, p. 417.
6. Kroeger, O. and Thuesen, J.M. (1989). *Type Talk—The 16 Personality Types that Determine How We Live, Love and Work*. New York: Dell, pp. 26–48.

8

The Propitious Time

Diego Centanni

CONTENTS

Isidora is the city of his dreams: with one difference. The dreamed-of city contained him as a young man; he arrives at Isidora in his old age. In the square there is the wall where the old men sit and watch the young go by; he is seated in a row with them. Desires are already memories.

The Invisible Cities by **Italo Calvino**

THE CONCEPT OF TIME

"Time" as a Symbol

Philosophers have argued about the nature of "time" for more than 2,000 years. When talking about time, confusion always arises between time itself and events, and the way we perceive these events, depending on the way "ourself" relates to them. Indeed, it is not time that flows, but rather events, and the way we experience them. The way that we perceive these events is a result of the interaction between the levels of our perception (conscious, pre-conscious and unconscious). This determines our experience of time.

As with space, time is not real or a thing on its own: it creates nothing. The meaning of time is given by the fact that it is a basic brick that allows the construct of society [1, p. xvi].

Within developed societies, clocks are one of the most important ways to represent time: but clocks are not the time. Time measurement achieved using a clock is, simply the measurement of the length of time, i.e., of the distance between two moments in time. In order to be measured, time is required to be temporarily stopped. We do so to create a model of time that allows us to describe time and have a shared understanding of the concept. The procedure used to let the movements appear immovable is a consequence of the total conscious objectification of the thought, necessary to transform temporal experience into a language that can be shared.

We can say that the clock reports the time, but it does so by continuously producing symbols that have a social shared meaning. Someone transmitting the phonetic model "time" expects that the receiver, if he or she belongs to a society using the same language, connects the same model of the memory to the shared phonetic model. This is the secret behind the communicative function of human symbols [2, p. 43]. When, during their development, the symbols reach a very high degree of conformity to reality, it becomes very difficult to distinguish between symbols and reality. Human groups that are not yet capable of regulating their affective and instinctive pulses by themselves, get help from the apparent hetero-obligation exercised by fantastic figures that reinforce their ability of self-control.

People belonging to a developed society, on the other hand, perceive the peculiar obligations of their character as the innate and conceive as a trait of human nature that makes them different from other people. Their imposed self-regulation according to the ever-moving time of clocks and calendars is a very good example—one among many—of the fact that both the obligations genetically determined by their own nature and the obligations of the social environment play an essential role in creating the individual habitat of a person. Each person is, to some extent, a guide for himself, as well as each person is, to some extent, subject to obligations that derive from life with others, and from the development and structure of the society to which he belongs [3, p. 42].

Socio-temporal obligation, which has now mostly become a self-obligation, is a significant model of the type of civilization-oriented obligations that are often peculiar to highly developed societies. But let's get back to time. Time, as a symbol, is a way of orientation that has been socially institutionalized. The mechanics of a clock, constantly moving in a given way, send a visual shared institutionalized message to each person, and each person is able to connect the right memory model to this visual model. In addition to the function of orientation, time has another function: it is a way to regulate human behavior. Time expresses the fact that in order to find an orientation, people try to determine positions, interval duration, speed of change, and many other aspects of the flow of time.

Chrónos versus Kairós

Chrónos is the chronological time, serial, measurable by clocks and chronometers. Kairós is the cyclical time, the time of episodes, human and living time of intentions and propositions. According to the etymology *chrónos* is the time as length, or interval: it is the time shown by the chronometers, by the clocks, which can be translated into a discontinuous sequence of dots over a line.

On the other hand, *kairós* is the time of human activity, opportunity, not of measurement. According to Greek mythology, *kairós* is the youngest son of Jupiter, the god of opportunities; *kairós* is related to his cousin, *kainós,* who is in turn linked to *kinein*, that is, the movement, which is the origin of *kinesis*. In short, the family of terms related to *kairós* is related with the time of movement, the time of the change, with the emergence of the new and active innovation.

John A. T. Robinson describes *chrónos* as "the abstract time, the time which flows objectively and impersonally, indifferent to what happens: it is the time measured by the chronometer, not by intention, which is transient and not significant"; on the other hand *kairós* is "the time considered in relationship to personal action, with reference to the objectives which have to be achieved in it" [4, p. 57]. Although *chrónos* has been handed down, through Latin, in every Roman language, *kairós* has remained anchored to classic Greek. This linguistic issue reflects the greater comfort, free of emotional implications, that we experience when we deal with a chronology.

In contrast, the time that emphasizes human intention and objectives and the consequent oscillations between success and failure, catastrophe and renovation, life and death, causes feelings of anxiety. In other words, *chrónos* and *kairós* refer to different ways of organizing the experience of time: time can be therefore defined as "the conceptual organization of the change experience," whose position and shape refer to our way of organizing the observation about what we think is going on in the world; every single person is the point of view from which the thought observing the world is originated [7, p. 55].

The Shape of Time

The concept of time, commonly used to read a clock and in natural sciences, is not adequate. The time of relativity is the time of the clock, used to date simultaneous points during physical events, but it does not have life and breath. The aim of this section is to explain the nature and shape of time, which is the relationship between vital processes and the flow of time. Understanding the nature of time means improving the knowledge of ourselves, of our actions, or our social relationships and of human life itself. The enigma of time is the enigma of life.

In concordance with our aim, we can say that time is a psychological phenomenon associated with the intention and psychic experience of events, rather than with the flowing of chronological time. Many of the barriers in understanding the meaning of time are in fact embedded in the enigma of past, present, and future. There is an idea about co-existence of the past, present, and future, according to which what is lived in the present is what has been lived in the past.

In this case the past takes the character of an evocation, and it acts as a constant pressure on the present, the pressure of what has been onto what is now. The past is not a temporal past: what is no longer and can only be remembered. It is something that exists here and now, and is present in the true sense of the word. In order to escape from the coercive influence of the past, it is necessary to cut off the bonds of the static concept of time. Time is dead if we limit our idea of it as driven by the movement of small cogs.

We live in a world mentally composed of action and change, and we live well until we try to understand its identity, talking about it. Problems arise when we try to describe it, both in poetry and novels, using either the precise mathematical language of physics, or the analytical language of philosophy. Indeed, in order to describe time it is necessary to do what words always do: to make objects immovable. Time is dynamic; words are static. Words seem useful to define objects, but they destroy the meaning of time just when they try to define it. Time of clock therefore defines events not as if they "entered into existence," but as simply existing, and therefore it admits that we "meet" them and we produce "the formality of their happening" entering into their future.

According to the anthropologist Benjamin Whorf whether two linguistic systems have a vocabulary and a grammar that are radically different, corresponding populations live completely different visions of the world [5, p. 55]. It can be said that even fundamental categories, such as space and time, are felt in a different way, depending on the linguistic schemes that constrain thoughts:

> The mental shape of an individual is controlled by categorical laws which act as a model, and the individual itself is not conscious of that. The thought itself is embedded in a language, each of which constitutes a vast system of models, each one different, inside which forms and categories are culturally ordered, through which the personality of the individual not only communicates, but also analyzes nature, observing or ignoring different types of connections and phenomena, channeling its own rational thought and building the frame of its consciousness.

According to Whorf, English sentences are composed in such a way that they show a given content or topic as part of an event, localized in a defined time and space. Differently, in the sentences of the Native American Hopi, events are not defined in relation to time, but rather in relation to the category of being, as opposed to becoming. The English language leads us to think about time as a divisible segment, which begins in the past, moves across the present, and continues in the future. On the other hand, Hopi grammar only distinguishes events that have already revealed themselves from the ones which are still undergoing.

Rhythm

How is it possible to get in touch with the environment, getting in tune with it? How is it possible, in an era of complete disconnection from nature, to find trust and naturalness? How can we ask more of ourselves than of others, leverage our own sense of balance, resist fads or the truth coming from others: follow our own pace, rhythm, without forcibly synchronizing with the rhythm of others. If we can successfully resist the environment, which would like to bind us at any rate and to our detriment, we will find the rhythm of our life also regaining peace and trust in ourselves. As with every true experience, at the beginning it is lived absolutely unconsciously: living in rhythm means to honor the truth and humbly believe in an ideal center, without which it is impossible to establish a real relationship with ourselves and the world around us.

Getting back to ourselves will also make us understand that the birth of our individual rhythm of life is indissolubly linked to the rhythm of the environment: because the time is formed when we confront ourselves with others [6, p. 18], and not withdraw into ourselves, facing the crisis of trust often experienced nowadays, could aim to a more balanced life.

Everyone lives in a personal time: two persons, in the very same moment, are not living in the same time. Each one, in the same moment, has his own personal temporal perspective. It might not be so evident that different people live different timescales. And the identification of the different timescales of the individuals is the key to solving many of the mysteries related to time and its nature.

SELF-KNOWLEDGE

Dream

If we analyze what happens when we dream, where time and space appear distorted, we realize that our conscious and conceptual reference point is still the notion of time. In our dreams, neither space nor time, as conceptual abstractions, are distorted, behaving in a strange way: what looks strange, in the conscious retrospective analysis, is the way the events occur: the end of a music performance can take place before its beginning, objects might behave oddly, an object can easily turn into another, maybe retaining the same appearance.

Therefore time can be used to control reality: the possibility of losing and altering the temporal experience, as well as the temporal destructuring processes of daydreams, as well as inside dreams or moments of difficulty of the self (e.g., schizophrenia), are all examples of that. Freud constructed an hypothesis about the nontemporality of the unconscious: "Nothing that corresponds to the idea of time can be found in the *Es*, no acknowledgement of the flowing of time, no alteration of the psychic process due to the flowing of time" [11]. The multiform reality of conscious, preconscious, and unconscious processes is in fact the way we build and experience the world, the way we act in the world and we react to it, with which we build our personal concepts of time and space [1, p. 67].

Oracle

In this sense, Oriental culture provides a fundamental help to Western culture, which could help to review its conviction about the robustness of modelization. "Everything flows like this river, relentless, night and day" [8]: this is how the idea of change is enunciated. Who has recognized the change no longer observes the single things flowing in front of him, but rather the eternal unchangeable law that acts within every change. Everything happening in the visible world is the externalization of an image, of an invisible idea. In this sense every earthly event is just a copy, so to speak, of an intangible event: everything is already present in every moment.

The *I Ching* (Book of Changes) continuously points out the importance and value of self-knowledge [8, p. 28]; it cannot be predicted, but it can be felt: stand still, be ready, resist the anxiety due to the waiting, resist the pressure coming from the customer and from other stakeholders who are satisfied only if something is produced.

The ancient Chinese point of view does not care about the attitude we put on when we get the oracle's response: we are the only ones who are puzzled, because we always stumble again in our prejudice, that is, in our notion of causality. Time and space are related to our sense of extension of the change, of the objects placed in reciprocal relationship, which change both in relations and in shape, in such a way that they are always either different or, which is the same, apparently remain the same.

For our practical purposes it is more useful to admit that everything changes, ceasing to cling to the deceptive security which comes from the belief that the objects can remain in exactly the same state or in the same place. Identity means continuity of existence and not static constancy and exactness of shape and place. Moreover, the number of possibilities for each evolutionary system is not predetermined: in the course of history, sets of possibilities merge, giving us the chance to reach opportunity which couldn't have been reached from the starting point. The new space of possibilities is not additive, but new sets of possibilities are generated. Every day we have moments when we blend together spaces of possibilities and we enter an unknown terrain: inside every choice we make, spaces of possibilities merge together.

From this viewpoint, it is therefore useless to anticipate the possible states: rather, we must find a mechanism flexible enough to generate constructive answers in the unpredictability of the incoming states.

My Self-Knowledge

Thoughts and sensations have been gathering in my mind for a while, concerning my perception of the world, for things that happen and will happen to me. There are some specific people with whom it is always the opportune time to talk about things that I do not even know where they come from, or why I always talk about them, I always have the impression that I am not telling these things to them, but rather to myself, maybe in the vivid hope of convincing myself that indeed they are not sensations, but just the way I perceive my reality which is simply different from the socially accepted way.

Thus I wonder about my perception, about my way of seeing and feeling the world in which I live. I also wonder what is time. Maybe I am the time, the way I cope with topics, the way I decide what I want to deal with: I could say that time is the way I manage my anxiety, my fears, my concerns, my dreams, and my feelings and emotions. In this perspective "manage the time" might be the same as managing my own anxiety, the fear of failing, of being inadequate, the fear of being judged in a way that is different from what I would like [9, p. 41].

We live in one of the possible worlds and maybe a world exists for everyone, because everyone perceives things in a different way, dreams in a different way, and contributes to shaping the world around him according to his model of a perfect world. So what can I do? How can I think that I know and interact with all this variety? I shouldn't do it. I only have to learn to co-exist with complexity and uncertainty; to work on myself, to work on my "idea of myself." So the development and the control of time of the project pass through the growth of awareness of the single members of the project team. I shall not shape people in a certain way, but I should route them toward their self-knowledge.

A person who has integrity and awareness is a person who attracts and motivates others to be the same. A person who becomes "transparent," is able to convey sensations and dreams and is capable of leading others toward objectives that are difficult, but somehow achievable.

What would I be if I were born in China? What would I believe to be real and possible if I had been educated in the East and not in the West; to what extent does what I do come from me, and what comes from the others? Which language would I use if I were born elsewhere, and had been educated elsewhere? I don't think there would be too much in common with the "me" of today. But it's still me: from the moment I was born to the moment I will die. So what? My reflections lead me to believe that everything comes from states of consciousness, maybe we are applying to our lives the same assumptions a priori that we use for "our" models, and abandon the opportunities that we believe we cannot seize.

We experience meetings, occasions, situations, and, sooner or later, if everything exists in every moment [8, p. 17], I can potentially encounter every opportunity. I can decide, whether I want to get those tools that allow me to be present, or just keep using the lenses provided by default by the "frame" inside which I've grown up, and according to whose rules I've been educated with. I am the only one who has the power to know myself; nobody can help me out.

What we often forget is that our mind experiences constraints that do not affect the mental life of others. Because we live within our culture, our brain is formed and governed by it; the ability of expressing ourselves through language does not automatically deliver freedom of thought, but rather it often traps us inside illusions and myths. We are the ones who have to fight to understand how our culture controls what happens inside our brains. Without this further level of awareness, the human mind cannot be defined as fully conscious. So what should we do? We stop and see, we work on ourselves to seize signs, and to perceive possible worlds, which today are unknown just because they are not perceived yet.

TOOLS AND ACTIONS

Free Up the Agenda

Do not act: "*Wu Wei er wu bu wei*: by doing nothing everything will be done" [10, p. 54 (Quote from "Tao Te Ching" of Lao Tzu, from the Chinese)]. Transform (*hua*) instead of acting: "nothing" is meant as *otium*, that is, as an active pause. Of course, I can do it if I can handle anxiety (anxiety of my boss, who overloads me with additional activities as soon as he thinks I am inactive, and my anxiety of filling in empty slots, as if they were a negative thing).

This is all due to the fact that being idle is socially perceived as negative, as if we were "losing time," as if we could have done more, or as if the success of my actions only depends on me. This is not the case. There are unknown factors (unknown to me, but known to others) that I do not consider in my strategic plans. They are real and they greatly influence the final result.

Time Is Not Just Linear

Perceive the project as many continuous and changing moments and not as an already planned sequence. Thinking of time as a sequence of instants that will never return simply creates anxiety. At the end we act just to mitigate anxiety. I feel inadequate because I do nothing, I do not use my time constructively enough. The time that counts is the cyclical one: you cannot oppose the linear one, so it is not worth being anxious over it.

Possible Co-Existing Worlds

To construct a project means to live in several co-existing worlds, possible worlds that interact. A meeting is a subworld, with its temporal rhythm. While we attend the meeting, we are living in other worlds as well: the world of our family, which continues even when we are not present, the world of my yoga class, and the world in which we have been and to which we will come back. And the same is true for all the others who live inside our world even when we do not perceive them as committed to it as we are.

Create a Non-Conventional Calendar, Based on "Rhythm"

Try not to suffer from the time of others. Time is one of the most intimate things we have. Being always available to others compromises the value of your own time. This will give meaning to your needs.

The Gantt chart is not the "chrono-program" of the project. It is just the idea of the project that I had the day I designed it. Tomorrow my idea of the project will be different. To remain constrained (which is also a consequence of the contract) to the initial program means to manage not the project but rather the variance with respect to the initial program.

Prevent Getting Worked Up

In a circular logic, delay does not exist. The never-ending, continuous doing, without pauses, always on, produces tension, both physical and psychical, subordination, and passive acceptance. To fill every moment is only good to allay anxiety and adds nothing to the objective. Live the time with moments of silence and uncertainty, of pause, which does not mean to do nothing, but rather to do something else. Governing time is to look to the near moment. The Gantt chart is just an output of what has already happened.

Redundancy

Invest in many structures that can rapidly answer to the needs of the environment to generate new functions.

Psychophysical Relaxation

Coaching training, autopoiesis, active listening, yoga, massages, spas, meditation (but also bike-riding holidays, trekking), anything that stimulates psychophysical relaxation brings serenity, availability, and empathy, the possibility to think away beyond ourselves, focusing on the possibilities and opportunities that we always encounter and that we are not able to be aware of most of the time.

Our societies and our offices are organized in just the opposite way, where everybody suffers, even mentally, from the anxiety of the others. Propitious time refers to a concept of "quality of time" and not only to a concept of "time duration." Time is episodic, made of circumstances, of glimpses of opportunities to seize in order to achieve a quality leap.

REFERENCES

1. Jacques, E. (1982). *The Form of Time*. New York: Crane Russak.
2. Elias, N. (1984). *Essay on Time [Uber die Zeit]*. Frankfurt and Main: Suhrkamp.
3. Berger, P.L. and Luckmann, T. (1966). *The Social Construction of Reality*. New York: Double.
4. Robinson, J.A.T. (1968). *In the End God*. Milan: Fontana.
5. Harris, M. (1987). *Cultural Anthropology*. New York: Harper & Row.
6. Schneider, M. (1965). Was ist Rhythmus? Uber die naturlichen rhythmischen Fahigkeiten des Menschen, in *Rhythmus*, XXXVIII, 2 pp. 18-24.
7. Maturana, H.R. and Varela, F.J. (1980). *Autopoiesis and Cognition, The Realization of the Living*. Holland, Reidel.
8. Wilhelm, R. (1991). *I Ching. Il libro dei Mutamenti [I Ging Das Buch der Wandlungen]*. Milan: Adelphi.
9. Rosenberg Colorni , E.R. (2006). *Lavorare senza offendersi*. Milan: Guerini.
10. Jullien, F. (2005). *Conferénce sur l'efficacité*. Paris, France: Presses Universitaires de France.
11. Freud, S. (1932). *Introduction to Psychoanalysis*. XI, 185. [Italian edition: *Introduzione alla psicoanalisi, Nuova serie di Cerini*, 1982, Opere, Torino, Italy, Boringhieri, 1967–1980.]

9

Leadership and Complexity

Stefano Morpurgo

CONTENTS

Melania's population renews itself: the participants in the dialogues die one by one and meanwhile those who will take their places are born, some in one role, some in another. When one changes role or abandons the square forever or makes his first entrance into it, there is a series of changes, until all the roles have been reassigned.

The Invisible Cities by Italo Calvino

LEADERSHIP

There are only three occurrences of the word "leadership" throughout the PMBoK®, third edition, which is indeed a sign of scarce interest in the topic. The situation was greatly improved with the publishing of the fourth edition, in December 2008, where the word occurrence grew to twenty-one, eight of which were concentrated in Appendix G, a new chapter dealing with "Interpersonal Skills." It goes without saying that the PMI felt the need to fill the gap.

According to PMI (PMBoK, fourth edition, page 417), leadership is defined as follows: "In general terms, leadership is the ability to get things done through others." On the same page, PMI provides further clarification of this interpretation of leadership: "Leadership involves focusing the effort of a group of people toward a common goal, enabling them to work as a team."

Such a vision of leadership perfectly matches the boundary conditions already seen in other chapters and typical of the PMBoK: clear objectives that the leader has to communicate to the team; full-time members to be forged until they become a cohesive team; and a directional, sharp communication flow in order to achieve maximum efficiency. But a further clarification should be carefully taken into account: "Respect and trust, rather than fear and submission, are key elements of effective leadership."

Reading between the lines, a kind of guilty feeling shows through: leadership cannot be asserted with the use of fear, let alone be imposed. PMBoK acknowledges that in everyday life a bidirectional communication flow is a value in itself, and that leadership is not a form of command, but rather the art of making people work together. These concepts are taken to the extreme when dealing with complex projects: objectives cannot be communicated top down, the project team is poorly defined, and everything co-evolves. In our projects, we constantly experience partial knowledge, a complex environment, and unpredictable events.

Indeed, every project carries its burden of complexity and even the smallest projects involve several people: customers, executors, users, controllers, and institutional members, all developing dozens, if not hundreds, of relationships, and ever-changing partially defined boundary conditions as the project evolves. The "assumption" of the PMBoK is to deal with linear projects. In the other chapters of this book we try to linearize the projects, to reduce their complexity. However, in this chapter we assume that complexity is intrinsic to a certain extent, and cannot be completely eliminated. Does it still make sense to talk about project management in this case? If it does, how should the project manager behave, inasmuch as he can neither foresee nor control his project completely?

WELTANSCHAUUNG

When facing a complex project, the first thing for a project manager to do is to turn his perspective around. Instead of either trying to foresee the evolution of an unpredictable system, or trying to handle boundary conditions that are out of his control, the approach should be: "OK, the system is unpredictable: How can I cope with it?" Such change in perspective leads to some fundamental implications that we will analyze briefly and call for new tools and techniques to be used in conjunction with the traditional ones.

An example can help understand such a qualitative leap: let's imagine the captain of a motorboat becoming a skipper on a sailing boat. He is used to precisely computing the course and fuel consumption, but now he has to radically change both his habits and his experience. He needs to cope with the prevailing winds, but he can neither forecast nor influence them. He would avoid, whenever possible, both the dead calm and the storm; and when he faces them, he would keep on sailing toward his destination anyway. He has to know his crew, but he really knows them only in time of need, when it's too late to change either the team or their attitude. Therefore he has to rely on each of them even if he does not know them deeply. He must have trained them to both the dullness of routine and the excitement and fear of unpredictable events.

After all it is nothing new; it is just what human beings have been doing for ages: survive in a hostile, unpredictable, unknown, and chaotic environment. The illusion of living in a linear predictable world has lasted just for a couple of centuries (a wink, compared to the whole history of mankind): indeed it is incredible that so much effort has been spent in order to devise tools and methods to handle small portions of simple projects. Nevertheless these efforts have not been completely wasted, because those tools and methods have given us the material wealth we enjoy nowadays ...

LEADERSHIP IN COMPLEXITY

Leadership in complexity is therefore much different from the leadership implied by the PMBoK, and can be summarized by the motto: "We are all ever ready for anything." Let's now see the implications of this statement. "We are all": there is no leader, but rather a plurality, or even a totality of leaders. The team must be able to act even if any of the members are absent: nobody is essential, nobody has the overall control, and everybody gives their contribution. If someone is missing, performance degradation might happen, but a complete paralysis will not.

"Ever ready": the group is constantly and continuously on guard, watching, searching, learning, and evaluating, relentlessly. Even at rest they are on the lookout, and the whole situation is not perceived as compulsory, but rather as a mission, a desire.

"For anything": everybody is aware that something unpredictable is always around the corner, that knowledge is always partial and faulty, that in spite of everything nothing is certain, and that the cohesiveness of the team is the only way to cope with the difficulties while being certain to succeed.

Therefore leaders (and we have just said that the project should feature a plurality of leaders) have to develop a sense and a love for the edge of chaos. They shun both excessive order and extreme chaos, and they do it consciously, constantly, and in a calculated way: they purposely introduce small disturbances or "provocations" when they perceive an excessive order, and they proactively tune it down whenever the project is about to go out of control. It is a dynamic balance, unstable, which should continuously be looked after, controlled, and adjusted.

At the same time, leaders pay constant and endless attention to their environment, which they explore relentlessly, looking for those feeble signs announcing the arrival of both the dead calm and the storm. They keep on exchanging information, formally and informally, about whatever they observe, and they collect whatever they find, "just in case." They are not only looking at the quantitative and explicit elements: the number of components of the project team, the expected duration of the project, the budget, the number of sites involved, and so on. And they do not only consider the quantitative but implicit ones, that is, those undeclared, like the overall number of stakeholders, the number of objectives, the acceptable amount of variation, and so on. They also browse through all the qualitative elements, such as the clarity of both the objectives and the constraints, the flexibility of the approach, the conflicts of interest, potential cultural clashes, and so on.

It is of the utmost importance that leaders explore the territory and collect as much information as possible about the presence and consistency of all these elements. Even if such knowledge is not helping to forecast the outcome of the project whatsoever, and it will not provide recipes or safe paths, nevertheless it will allow a much more conscious management and it will reveal all those feeble signs which, if ignored, can cause a project to fail.

WHERE DOES COMPLEXITY LIVE INSIDE A PROJECT?

Within living systems, and human ones are no exception, complexity arises from the relationships that exist between members. Within human systems, and projects are no exception, relationships are represented by communication between persons. In complex systems, communication management is therefore one of the main tasks of a leader who will constantly strive to maintain the system on the edge of chaos [1]

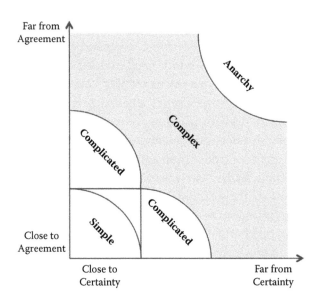

FIGURE 9.1

The edge of chaos. (From Stacey, R.D. Strategic Management and Organizational Dynamics. *The Challenge of Complexity.* Prentice Hall, 2000.)

(see Figure 9.1), to keep communication alive and unpredictable, although suitable for the context to which the project belongs.

These concepts are nothing new, and they apply to several examples: during scouting and analysis phases communication would flow freely "from many to many," as well as during a brainstorming session; during operational and executing phases in controlled environments, communication would fall top-down in the form of orders or recipes to be executed; and so on. Successful leaders instinctively adapt their communication style to the present needs, thus allowing a smooth evolution of the project.

A further element of complexity arises whenever the project is large and subteams develop inside the project itself, and therefore relationships exist not only between individuals, but also between groups. Such an additional element of complexity can be experienced more often than one would be willing to admit: in fact there is a tendency to look at the project team (no matter how small and somewhat controllable) rather than at the whole set of players who interact with the project, either directly or indirectly. Underestimation of how important it is to handle relationships between all existing groups is one of the main causes of issues for a project.

Management of relationships between groups calls for skills that are more sophisticated than those needed to handle relationships between individuals, and it also calls for different tools, as we show later in this book. At this point it is enough to remember that project managers in complex environments should take care of communications between both individuals and groups.

WHEN PROJECT MANAGEMENT TOOLS SHOULD "NOT" BE USED

The reader is probably wondering: "What should I do with all the nice things about project management that I've studied thus far? After all, they helped me to manage several projects successfully." Well, these techniques are still valid to manage all those "linear stages" that occur inside a complex project. It is worth repeating once again, no matter how complex the project as a whole, nevertheless there are stages and periods linear enough to be managed as a "simple" project. Recalling the skipper example, the course is a zigzag of straight courses, each of which can be, and should be, managed linearly, because applying the complexity paradigm would result in a waste of energy. Similarly, managing the whole course as if it were linear would lead to a complete disaster.

TOOLS FOR COMPLEX PROJECT MANAGEMENT

It is time to go over a set of tools project managers can use to manage a project in a complex environment. It should now be clear that none of these tools is a magic wand able to calm down a hurricane, because hurricanes cannot be eliminated completely. Moreover, these tools are not meant to be mutually exclusive, but indeed the right mix should be used. Finally, no result is guaranteed a priori. Exploiting the right blend of experience and humility, smart project managers eventually learn how to properly select the right mix of tools to achieve the highest likelihood of project success.

Polyarchy

"We assume leadership is something done by the (we hope) talented few, exercised over the (presumably not so talented) many. [...] Polyarchy [...]

means leadership done by the many" [2, p. 2]. I cheerfully welcome the term "Polyarchy" in lieu of either "cooperative leadership" or "diffused leadership" because the term "leader" and its derivatives always implies the idea of "command" or "guidance," and it often stimulates questions such as, "What are the features of a good leader?" Polyarchy is more neutral in that sense, and it immediately shifts the emphasis from single to "many."

In several cases, a centralized management is inefficient, when not detrimental, even if the task to be carried out is a simple one. As an example, let's consider the movie clip where a group of about 30 people have to line up in such a way that each one is equally spaced between two other randomly chosen people [3]. By letting people act autonomously, the group is able to get to the final arrangement in less than one minute. What would have happened if someone had been put in charge? The overall time to solve the problem would have been much longer. In this kind of situation, polyarchy is the most effective approach. The diagram shown in Figure 9.2 depicts those circumstances where polyarchy [4] is effective.

What then is the role of a project manager inside a polyarchy? The project manager, like the gentleman who provides initial instructions in the

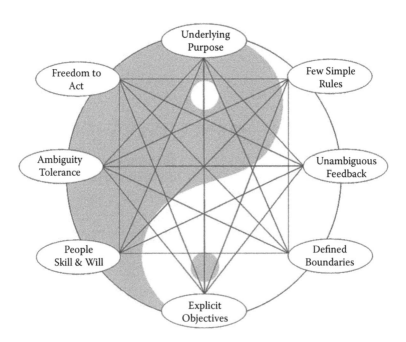

FIGURE 9.2
Polyarchy. (From Obolensky, N. *Chaos Leadership and Polyarchy—Countering Leadership Stress?* University of Exeter, 2007.)

movie clip, is the caretaker of the situational features described in the diagram, carrying out several tasks: to validate such features, to show them to the participants, to make them fully understood, and to encourage both perseverance and respect among team members. The project manager enables polyarchy whenever appropriate, collecting the result of teamwork, commenting on it, and giving it back to the team.

Polyarchy is a powerful tool, and it is capable of generating results so outstanding that they just go beyond team members' expectations. On the other hand, polyarchy can lead to extremely chaotic outcomes and even make a project fail (no matter how genuine the effort, the good faith, and the good will of the participants) when either misused or improperly managed. As with any powerful tool, it should be handled with great care.

Neighborhood and Ethics

Indeed, it is easier said than done. In a complex project, instability areas can arise, and they can have wicked and wide effects when underestimated and neglected. Smart project managers must "hang out in the neighborhood" of any critical situation, so that it is easy to spot the initial spark and put it out before it goes out of control. Danger is not something to be afraid of, but rather something for which we actively search. Proactive listening is one of the best methods to be a good neighbor.

This attitude can be an example for the rest of the team, which becomes more conscious and proactive in the management of criticalities. The project manager should be concerned (one would say "slightly paranoid") even in the opposite situation, when everything is apparently going well, and everything is calm, too calm. As already pointed out, being on the edge of chaos is the name of the game.

Ethical behavior is a close relative of being a good neighbor, and it is as effective in crisis prevention. The topic is not discussed in detail: suffice it to say that the aim of ethics is to eliminate any friction among team members, in order to prevent both an unnecessary waste of energy and a serious threat to project completion.

Cosmological Principle

Edwin Powell Hubble was an American astronomer who discovered that the other galaxies apparently move away from our own, which appeared to be the center of the universe. Hubble firmly turned down this hypothesis and embraced the "cosmological principle," which states that in the universe there are no privileged observation points. Therefore the same theory applies throughout. As a consequence, Hubble concluded that the galaxies all move away from each other, and therefore the known universe is expanding.

This principle is applicable to complex systems as well, and the implication is twofold: no privileged observation points exist, and no privileged roles exist. It is now important to fully understand the meaning of such a twofold principle, and above all the benefits for a project manager to use it. Roles can be considered first, and that is the easy part. A project is a complex system where any attempt to favor one role at the expense of another has always resulted in injuries to both. A relationship with ethics exists, but to a wider extent, involving both roles and groups. The task of the project manager is to make sure that all the players are conscious of the importance of the Hubble principle, actively looking for an overall balance between roles. It is worth noting that the project manager cannot explicitly enforce the balance; according to the principle, all members have the same rights and duty. It is a subtle blend of politics, diplomacy, and persuasion.

Observation points are much more important and harder to explain. The project is not only complex in itself, but it is also tightly interacting with a complex environment, both externally and internally. An essential aspect of interaction deals with the creation of images and models of the environment with which the project interacts. A model is, by definition, an approximation, far from being perfect. And the higher the number of people contributing to the model, the better; in order to maximize the completeness of the model, everyone must contribute, no one excluded. Again, the task of the project manager is not to direct the effort, but rather to foster the principle. In this way, all the players are led to embrace it, maintain it, and apply it. All the points of view are then accepted and respected, appreciating the fact that each is a complement to his or her own. The Bantu language has a specific word, *Ubuntu*, which synthesizes this concept: it is much more efficient than English, where a whole page is needed in order to explain such an important thing.

Convergence and Divergence

Every project evolves through alternating divergence and convergence phases, through analysis and synthesis. A smart project manager is able to "tune in" this wave and seamlessly guide the project throughout the phases. It is a common belief that the duration of the phases is either fixed or can be determined in advance, but unfortunately reality is different: trying to force an oscillation of a system according to a frequency different from its own can lead to a complete disaster. When dealing with this topic, the project manager needs to master the same skills as a disk jockey in order to give the project the right rhythm to dance to according to the music. The signs are feeble, but a good ear can catch them. For instance, let's consider the analysis of the requirements for a given topic: at the beginning, the team acquires new information at a very high rate. At a certain point the rate gets to a plateau, then it begins to drop. Reasons are manifold: the framework is already clear enough, the team is just tired of brainstorming, and hypotheses need validation before a new round of analysis.

If the rate decreases, it is time for a phase change. "When" it is going to happen is just unpredictable, inasmuch as it depends on too many internal and external factors. Again, the sensitivity of the project manager is crucial. It is important to point out that, unlike what happens with simple projects, the achievement of a result in a complex project should not be used as a target time to switch phases: most often, it would be a big mistake.

Control Parameters

It should be clear by now that a complex system cannot be predetermined. It evolves over its own paths, and the project manager can just try to guide it. The best driving tools are the so-called "control parameters" [5], that is, a set of boundary factors which modify the environment the project moves in, and are able to influence (but not to determine) project evolution, narrowing the cone of potential paths. For instance, the hotter the climate is, the higher the number of accidents, the higher the number of mistakes, and the lower the productivity is. Therefore even temperature is one of the many control parameters that the project manager needs to consider, because it influences project evolution.

On top of environmental factors, another fundamental control parameter that the project manager must always take into account (in his role of communication manager) is the level of communication. Just as the success of a party can be judged by the noise (and an experienced host keeps it to an acceptable level, neither too loud nor too low), so the good health of a project can be inferred from the fizz of communications. The aim of the project manager is to keep it far from the extremes—deathly hush and street fights—using several techniques: workspaces are arranged properly, both formal and informal gatherings are scheduled, witty and confrontational questions are sprinkled around, tones and pitches of conversation are always carefully monitored. For instance, the amount of e-mail exchanges should be constantly controlled as well. Ethics, meant as the achievement of mutual respect, can be also considered one of the best control parameters available to the project manager.

Priming and Mindfulness

The category of control parameters also includes two somewhat complementary tools: priming and mindfulness. Priming refers to the ability that a stimulus has in facilitating a response by the receiver. Response, as is always the case with complex systems, is not automatic, but is more likely thanks to priming. For instance, reading the same text written on a red sheet rather than on a white one triggers different behaviors (the red sheet conveys a sense of fear and alert). For one example, the way a question is asked can lead to different types of answers.

Mindfulness is the capability of identifying the priming: awareness or consciousness. It means to pay attention to the environment around us, to be able to understand the reactions (of others and our own) in such an extended context. For example, the defect rate is influenced by the day of the week, by the time of day, by the weather, and so on.

Priming and mindfulness can be used together: studies show that productivity and harmony can be improved by some types of music, proper lighting, appropriate furniture layout, and the like. A good practice of mindfulness and a wise use of priming, endorsed by the whole workgroup, are a great opportunity of boosting the project team's performance.

Models and Simulations

Any complex system can be better studied and analyzed building a model and simulating its behavior. In some cases, including co-evolutionary systems, a model is the only way to make a reliable forecast. Because building a complete model is often too expensive and time-consuming, sometimes it is more convenient to build a partial one, which then interacts with actual components of the system. For instance, information systems that modify the way users work can be partially simulated and then provided to the real users, who act as beta-testers.

Using models and simulations is a great way of embracing project complexity, because participants can focus on results, rather than on the way of achieving them. Furthermore it provides a full-picture view, instead of concentrating on the details. Last but not least, project flaws can easily be spotted in advance and fixed: indeed, a useful tool, alas scarcely used.

Role play is a modeling and simulation technique: actual system scenarios are simulated using different sets of organizational, communication, and environmental models in order to verify the resulting outcomes. Another useful tool (described later) is catastrophe simulation.

Self-Organization

Complex systems approaching the edge of chaos exhibit a tendency to self-organize. This feature is neither good nor bad in itself: self-organization can either be of help in achieving the project goal or completely counterproductive. The important fact is that once started, self-organization is difficult to be either stopped or modified. Leaders should therefore pay great attention during the initial phase of the project, when self-organization is more likely to arise, in order to promote those forms that are really functional to the objectives. All the tools described in this chapter can be used.

The following empirical rules can help in setting up a smart form of self-organization [6]:

- A set of simple rules leads to complex and smart behavior. A set of complex rules leads to simple and dumb behavior.
- Self-learning leads to smart self-organizations. To promote self-learning means to encourage trials, to tolerate errors, and to provide feedback on achieved results.
- Operational space has to be bounded, but innovative behavior and creativity should be able to grow within the set limits.
- Information is not conservative: whenever information exchange is supported, the overall informative wealth of the system is enhanced.

Redundancy and Scarcity

When dealing with complex situations, it is important to be able to cope with uncertainty, reduced communication capability, and limited resources. This ability should not be regarded as a theoretical exception to the rule, but indeed must be a fundamental part of the skillset of the participants. Leaders should stimulate people to take the initiative, to be creative and solve sudden unexpected issues (to think out of the box, so to speak), and to make the most out of available resources, no matter how scarce they might be. Rather than spend time in carefully allocating resources, it is more rewarding to develop creative thinking which helps in using available resources effectively. It is of the utmost importance that redundancy exists within the organization: that is, backup resources to be used in case of need. Scarcity is, to a certain extent, a complementary asset, inasmuch as the team must be able to do more with less: balance on the edge of chaos, reloaded.

Spontaneous and Controlled Perturbations

Behavior of complex systems cannot be inferred from the behavior of their components. The easiest way to collect information about complex systems is to trigger small perturbations and observe their development. For instance, the adaptive capability of ants searching for food can be studied by placing obstacles on their path and observing their behavior. This approach is much simpler than thoroughly studying a single ant and then inferring the overall behavior of the whole colony. Project managers can efficiently use the same technique.

Small perturbations often occur during the project life cycle, and they can all be viewed as case studies. Changes in team composition, in project goals, in available resources, in boundary conditions: countless events constantly perturb the project. In our new Weltanschauung, these are all potential learning opportunities, because they reveal the ability of the team to react to unexpected events, the tolerance of the customer to variations, the level of environmental stability, and so on. Leaders grow their knowledge each time they study the evolution of the project under all these perturbations, and they learn to sail their boat even in heavy seas.

Perturbations can be purposely introduced by either the project manager or the project team. For instance, a subsystem can be deliberately switched off, an activity can be stopped, an objective can be modified, and a resource can be deallocated in order to observe the consequences. Controlled perturbations are learning opportunities that are even more interesting than spontaneous ones. If the whole team agrees to use them, then the resulting benefit grows exponentially. On the other hand, complex systems are unpredictable, therefore introducing a perturbation means to be ready for anything!

Catastrophes

Complex systems are catastrophe-prone, and projects are no exception. Contrary to the common use of the word, in this context catastrophes are not necessarily negative events for a project: an expert leader can "ride" them to meet the objectives. Let's now analyze some kinds of catastrophes and their possible usage.

External natural catastrophes such as fires, floods, and the like are the most dreaded, a mix of rational and atavistic fear. They cannot be prevented completely, but their consequences can be mitigated by means of insurance, disaster recovery plans, and above all simulations (as mentioned in previous sections). Natural catastrophes usually hit large communities and they tend to affect the weakest members while strengthening the survivors. They play a fundamental role in the evolution of the species, and in general in the evolution of complex systems, including the subjects of this book. Wise management of these events can be a competitive advantage that would not be achieved otherwise.

External artificial catastrophes including failures, budget resets, takeovers, and so on are not as frightening, and therefore less analyzed. Little or no countermeasures are usually taken, thus artificial catastrophes have tougher consequences than natural ones when they occur. They generate uncertainty, against which paralysis is a typical response: and a paralyzed project does not survive. The winning strategy (it should be self-evident by now) is prevention, that is, the ability to overcome uncertainty and complexity by capturing in time the premonitory signs; to keep moving on the edge of chaos neither freezing nor blowing up the project, following the course despite the adverse conditions, belt-tightening to save resources; to consolidate the neighborhood by tightly controlling ethics, which is always put to the test when the going gets tough. Last but not least, a team who has simulated similar events is stronger, less anxious, and much more aware of its residual resources.

Internal catastrophes include mutiny, strikes, resignations, internal conflicts, and disputes. All these events are well known to experienced managers, therefore it is not worth dwelling on them. It should be noted that for the most part they are premeditated actions executed to wipe out a form of self-organization that is no longer suitable for the system, in order to replace it with a new one. In fact, because self-organizations feature mechanisms of self-balance to keep their stability, the only way to move from one to another is through a catastrophe. Not necessarily a bloody one, it can be more subtle, but in any case the result is an abrupt change of internal balance. Leaders must be aware of the fact that some situations might evolve in sudden leaps rather than linearly, and should therefore act accordingly.

HOW TO TRAIN A "COMPLEX" PROJECT MANAGER

As already pointed out, leadership in complex environments is a well-balanced combination of skills, knowledge, attitude, and tools. How can a project manager achieve the right mix? To begin with, it is important to understand that all these features are neither alternative to nor conflicting with the classic ones; they are in addition to them. A project manager needs to master both the skills and the know-how described by the PMBoK before getting to the next level. Then, as experience grows, a smart project manager realizes that the classic tools have limits, no matter how correctly they are used. The desire to overcome such limits leads to the search for complementary methods.

Reading this book is a way to lay the theoretical foundations. But how to really manage complexity can only be learned on the job. Everyday life is a great school, and even better when mentored by an experienced project manager who has already endorsed the Weltanschauung mentioned before. It is the same approach followed in the schools of Oriental philosophy: love and respect for complexity is better gained living with the teachers.

Another learning approach, though a less conventional one, is to use role games and simulators, the so-called massive(ly) multiplayer online role-playing games (MMORPG, for short): computer role-playing games in which a very large number of players interact with one another within a virtual game world. Thousands of players assume the role of fictional characters, interact, and evolve together with the fantasy world in which they live virtually.

Fans of this kind of game are becoming a matter of study for both researcher and talent-hunters. An interesting paper published by the *Harvard Business Review* [7, pp. 58–67] explains that these games require (and develop) highly desirable skills: to select, to involve, to motivate, to compensate, and to retain people from another culture and background;

to identify and leverage competitive aspects of an organization; to analyze several sources of incomplete and ever-changing data; to make decisions rapidly whose consequences might be both pervading and long-lasting. Online games take these skills to the limit, because organizations are "composed of a volunteer workforce in a fluid and digitally mediated environment."

In such environments, it is not easy to understand how leadership works, nor who the talents are: it takes some time to learn the game rules, and it takes hundreds of hours of gaming to gain the competence level that allows us to appreciate fully the skills of the most experienced players. IBM commissioned a research company to study leadership in games; for eight months, a team of a half-dozen veteran players with more than 50,000 hours of cumulative experience was observed, and the actions of leaders were recorded. A follow-up survey at IBM of people with both gaming and business leadership experience suggested how the same competences are shared between gaming and real-world corporate contexts.

As a conclusion, this chapter highlights the fact that environmental factors are important to enhance the leadership skills for which we are looking. Such factors can be peculiar to the gaming world, but it is a strong belief of the authors that they could be adopted in real-world contexts, where they would act as a catalyst to form a new class of leaders.

AN EXAMPLE FROM ITALY

We would like to end this chapter reporting some sentences from the speech by Sergio Marchionne (CEO of FIAT and Chrysler Group) on receiving an honorary degree from the Polytechnic of Turin. Many of the concepts discussed in this chapter can be found there, and it is encouraging to see that they have been successfully applied in an Italian industrial context leading to great results that went beyond the most optimistic expectations of company management.

> We are building a new FIAT on the roots of the old one. [...]
> The world we operate in is complex, chaotic at times. The problems we have to face are different every day. Variables at stake are so many and so great. Because of that, the system must be as flexible as possible. [...]
> It is not easy for a company so large and complex like ours, but it is mandatory if we want to seize all the opportunities. And it makes the difference between win and lose.

In such framework, the only things we can set are the objectives.

We did it clearly and rigorously, in spite of market conditions and global economy. Maybe that's the reason why someone was so surprised when we achieved them [...]

And yet we did not change our objectives [...]

Now our commitment is to make sure that the change is no longer something we undergo, but rather something we naturally deal with.

A common approach exists, which generally applies and which helps solving even the most complex issues.

That is, to recognize the central role of both the people and the leaders who manage them. [...]

That's why it is so important to have good managers, who are cohesive and who share the same direction, the same methods, and the same objectives.

It is the unity of all the leaders around a set of shared value. [...]

By "smart leaders" I mean someone who has the courage to challenge the Obvious, to follow unexplored paths, to break with tradition and old habits, to think out of the box.

Men and women who comprehend the concepts of service, community, and respect for the others.

People who act quickly, but who are also good listeners, and highly reliable. [...]

We must grant them both freedom of action and decision power. We must provide a merit system, because it is the only way to secure the best talents for our Group and because it is the best way to let everyone emerge and show their value.

We must give them the opportunity to grow, because it is the only way to assure the growth of the company as well. [...]

But what I am most proud of is to witness their success and the positive impact on them, on their self-confidence, and on the vision they have on their own future. [8]

CONCLUSIONS

The words of Sergio Marchionne just confirm the concepts stated in this chapter, which can be summarized as follows. A project is a complex system that survives only if it constantly moves on the edge of chaos.

It is a joint effort, where all participating members have an active part and are all leaders, striving to achieve the objectives. Objectives are a lighthouse to be reached. But the course is not known in advance: it is built daily by everybody, using both the planned resources and those that "by chance" are gathered along the way. To be successful, the gathering activity should be continuously performed by everybody. Nothing and nobody should be neglected, forgotten, or thrown away, "just in case." Even what today seems useless, or even dangerous, like a catastrophe, could ultimately turn out as an advantage one day.

REFERENCES

1. Stacey, R.D. (2000). Adapted from *Strategic Management and Organisational Dynamics, The Challenge of Complexity,* 3rd ed. Upper Saddle River, NJ: Prentice Hall. www.plexusinstitute.com/edgeware/archive/think/main_aides3.html and www.gptraining.net/training/communication_skills/consultation/equipoise/complexity/stacey.htm (accessed January 8, 2012).
2. Obolensky, N. (2007). *Chaos Leadership and Polyarchy—Countering Leadership Stress?* (Extended Essays Series, Center for Leadership Studies, University of Exter).
3. See http://www.youtube.com/watch?v=41QKeKQ2O3E.
4. See also www.complexadaptiveleadership.com/images/ChaosLeadership.pdf, p. 2.
5. Giancotti, F. and Shaharabani, Y. (2008). *Leadership agile nella complessità. Organizzazioni, stormi da combattimento [Agile Leadership in Complexity. Organizations, Fight Formations].* Milan: Guerini.
6. Dubakov, M. (2009). Extracted from *Simple Rules, Complex Systems and Software Development* (see http://www.targetprocess.com/blog/2009/03/simple-rules-complex-systems-and.html).
7. Malone, T.W. and O'Driscoll, T. (2008). Leadership's online labs by Byron Reeves. *Harvard Business Review,* May.
8. Marchionne, Sergio. (2008). Speech to Turin Polytechnic. www.polito.it/ateneo/grandi_eventi/lauree/marchionne/marchionne.pdf

10

Narrating to Believe

Bice Dellarciprete and Andrea Pinnola

CONTENTS

As this wave from memories flows in, the city soaks up like a sponge and expands. A description of Zaira as it is today should contain all Zaira's past. The city, however, does not tell its past, but contains it like the lines of a hand, written in the corners of the streets, the gratings of the windows, the banisters of the steps, the antennae of the lightning rods, the poles of the flags, every segment marked in turn with scratches, indentations, scrolls.

The Invisible Cities by **Italo Calvino**

PR(E/O)MISE

"You have just been assigned to a project which already started and you want to quickly understand where the project stands. Where are you going to look for information?" "People coming from four different countries work on your project. How do you manage communication?" These are typical questions from a PMP® examination, therefore a "right" answer exists.

Planning project communication, giving it a structure, formalizing it, and managing it are covered in a book of etiquette for these tasks that is detailed, elaborated, and geographically diversified. We have studied the PMBoK®, and many more. Nevertheless, we sometimes feel that something is not under our full control. We often wonder if, on the spot, we can really say where the project stands, or whether the team members have the

FIGURE 10.1
Comprehending the project.

right tools and occasions to communicate and co-operate, or how important our activity is within the hosting organization.

The context develops over time, and it evolves. New stakeholders come and others go. The path is always in progress. The need for comprehending the project as a whole, as a dynamic process, and from different points of view gets more and more relevant (see Figure 10.1). Formal reports, progress memos, and the like represent what has happened and what might happen for the benefit of the specific stakeholders to whom they are addressed. Experience teaches us to be cautious when we state the actual progress based upon the hypothetical one, or when we look for lessons to learn, or when we set the premises for a project with similar scope.

The project manager as "communicator" should also be able to get attention at the right time from the right people. Please see Figure 10.2. Quite often, the involvement of the customer, of the sponsor, of the project members' functional managers is suddenly and unpredictably needed. In particular, a great deal of the project manager's role is to create a team where individual contributions sum up (at least) instead of clashing and getting dispersed. We all know that it is often more important to rely on the understanding of the context and on the willingness to co-operate rather than on the specific abilities of the single members.

FIGURE 10.2
Orchestrating without a musical score.

Participation in the journey of the Complexnauts has given us the opportunity to "serenely" think about these topics. We have wondered which tools (in addition to the ones indicated by the PMBoK) can be used in order to both interpret and represent the different facets of our work that cannot be framed into predefined, predictable, prescribed schemes. Beyond the prescribed … a white page, maybe, but not only that.

Storytelling has already become a fad, proclaiming motivation, understanding of the change, generation of consensus, showing off of achieved results, and knowledge-sharing among others. Common talks among project managers include project tales and related recipes, questionnaires, templates, standard topics, and post-mortem rituals. Something is still not convincing us, but the idea of "narrating the project" still looks promising and appealing.

Indeed it is useful to create and collect project tales, as well as collect unofficial documents and simple evidence of the facts, making this material available, analyzing it to spot both emergences and relevant details, in order to understand the worlds that originated the collected traces. In this way it is possible to trace plots that resemble the "real" project, representing both facts and people involved in its evolution, and use them whenever appropriate for several different audiences.

The idea that inspired our journey is the same metaphor adopted by the Complexnauts crew: to build and maintain a memory of whatever occurs during the project lifetime—a rich set of facts and experiences—in order to lower the center of gravity (thus increasing stability) of our acrobat who is walking on a tightrope over the edge of chaos. For whoever walks in uncertain territories, to know and to be able to recount his own history is a source of stability (balance and ability to rise again after falling). This is true for the project, for the individual working on the project, and for the hosting organization. In other words, a wealth that should be capitalized and managed in order to cope with complexity.

This is a nice thing to be said, but is it achievable? Isn't it too much to be expected from a project manager? The project manager (see Figure 10.3), already busy for the 80% of the time in "traditional" communication and for another 100% aiming at a moving scope, balancing schedule, cost, and quality, while the sponsor is sleeping, the customer wakes up at the wrong time, and the team swerves in uncertainty. Should the project manager recount, too? Is it a joke?

Indeed, it is not "another job." Rather, it is a higher gear to shift to whenever a project faces uncertain, changeable, hostile, or geographically

FIGURE 10.3
Project manager as a storyteller.

fragmented situations. We have investigated further, mulling over things we have experienced and things that we would like to live to tell. Here are our travel notes, including some initial encouraging discoveries.

"I'VE SEEN THINGS"

When we started to confront ourselves with the Complexnauts crew, we noticed a peculiar way of narrating: "I (the project manager), can write some stories (because I like doing it, because it is congenial to me, because I want to communicate, to involve people, to stand out, etc.)." Along the way, we have considered more and more the importance of other topics: to allow the emergence of stories recounted by others, to consider and represent the viewpoints of several interlocutors, to record events that have been neglected by the "official version" of the facts, and so on. In general: this entails dedicating more time and attention to collecting what really

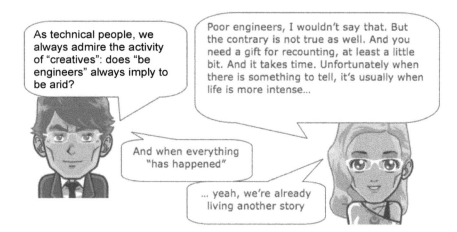

FIGURE 10.4
Observing and listening to the project's participants.

happens on the project, observing and listening (see Figure 10.4) to the context in its variety and complexity, and even more, to pay attention and wonder what the project represents (and how it can be represented) for its various stakeholders. By the way, stakeholder is used as a synonym for "person interested in the project": but indeed the project might be either in conflict or totally irrelevant to the stakeholder's interest.

How can we stimulate proper attention? How can we recount it to them? We believe that in many cases it is worth investing in a "narrative" approach. Our belief is based on the things we have done, on the things we could have done better, and on the things that others have recounted in a persuasive way.

Throughout our career, we have often reopened the archives of accomplished projects, looking for a certain document, as if we were rummaging in an old trunk. By chance we came across an e-mail or an informal note that condensed in a few lines the true story of the whole project, better than thousands of properly formatted pages. (Did I read it? Did I really understand it? Why didn't I consider it important? What did I consider important instead? Did I replan justifying project deviations with respect to an old Gantt diagram?)

Sometimes we have received the documentation of projects handled by others, often with a sense of boredom and diffidence, sometimes in an investigative mood, learning to find and interpret errors and omissions above all. We have experienced frustration when we have seen collaboration tools and other sophisticated mechanisms installed in our premises which then proved to be unsuccessful in our own projects. And when we tried to execute "with quality" but succeeding just "to some extent" (the project manager and the quality auditor—the thief and the policeman—we performed both roles at the same time).

We have handled innovative and complex projects trying to believe in "the sensitivity of the project manager to the important details," achieving unpredictable results. But how much time did we waste specifying models of future projects (then executed in other ways), detailed, quantified, shared, validated, and frozen in our perfect plans. A last, but not least, intimate note, we often wonder: "What will we keep, out of such a huge job?"

Let's Play

So, we understood that we were sharing a certain way of seeing things. In order to present our first ideas to our Complexnaut colleagues properly, we tried to settle on a method.

> Dear Bice, after our conversation I have written down some ideas considering some hints that I found in the article you sent me. You can find everything in the attached ppt file. My idea is to draw a mind map of the "project narration" to be used as a facilitating mechanism, but we can also think to something else (I would have liked to prepare a sort of narrating presentation). Cheers, Andrea.

> Dear Andrea, what a gorgeous day. I got to the office unusually happy [...] And relying on the fact that I have time to keep on re-organizing my notes. And now I read your email. Well: I am sketching a map too, using—figure this—the various who, why, where, when, etc. as main nodes. ... Looking at your map it seems I have something to add. I'll see how to do and get back to you. Bye, Bice

That's how our first "map of narration" was born. Narration, we thought, unfolds around "who" narrates. Moreover the focus on "who" narrates is fundamental because it allows the emergence of those characters obscured by the official version of the events. We considered first of all the various stakeholders that had something to say in the project. "Who?": the project manager, the sponsor, the team, the customers, the suppliers, the performing organization, many people who either consciously or not live to tell stories that get tangled, forming the context of the project.

Then we wondered "what" constitutes the narration of the story of the project. The scope, of course, and whatever is usually documented in the plan. But that's not enough. Details that matter can hide everywhere. Things written on the map that were apparently irrelevant have played a fundamental role in the history of some of our projects.

Studying the collection of our first ideas, taken as sticky notes on a wall, helped us a lot to find various, interesting, unpredictable, and absolutely meaningful cases of "who narrates what." Encouraged by that (see Figure 10.5), we went on wondering: "why?" can I narrate and therefore: "where?", "how?", "when?", "how much?"... and lastly "to whom?" We got carried away, so we have also computed the possible combinations of these elements, throwing numbers, 8! = 40,320, a dense plot of possible paths to try (see Figure 10.6).

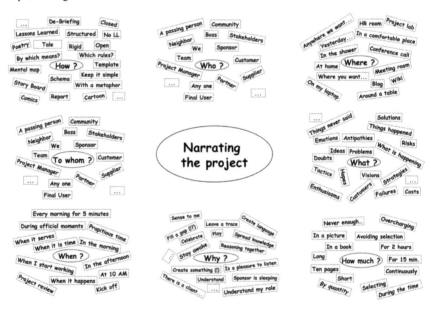

FIGURE 10.5
The narration map.

Try it yourself: create a sheet for each question (Who narrates? What? etc.) and write all the answers that come to your mind, or take the map we suggest and customize it both adding and removing items. That's interesting, isn't it?

FIGURE 10.6
Customizing the map.

EASIER SAID THAN DONE

Our first map is a little bit naïve. We have shown it because it helps in understanding the kind of everyday experience we have while writing these pages. We work in information technology (IT), in particular, we deal with web projects, under the different meanings that this word got during the years: interfaces for service access, collaboration tools, and Web 2.0, among others.

Our job aims to invent, conceptualize, and translate into functional and technical terms, and create "something" which, depending on the point of view of different stakeholders, can be described in very different ways. Can we apply what we are saying to many (all the) projects? Maybe. Making comparisons with other facets of the IT domain (for instance, the ones where transactional systems are designed), the product itself of our projects can be considered particularly "complex" (a simple example, known to everybody: all the fine details that compose a web page to increase its usability). Furthermore we live in a world where an incredibly high number of stakeholders are present, together with their languages: system administrators, web designers, graphical designers, Java programmers, content providers, and on and on.

In these situations, experience has led us to take with great care the heterogeneity of the worlds represented by the various stakeholders, in order to try to understand their extremely different ways of relating to the project. Therefore we must pay attention to the details that reveal such differences, which also show through the way they act and communicate, or even hide and shut up. "Not being there" is also an important signal, and so is allowing or urging informal expression or "in first person" of their expectations, competences, complaints, and concerns. We must commit to the comprehension of the quantity of fragmented information, of the additional "disorder" that was created for this reason. We have tried to adopt this approach in different possible situations, compatibly with organizational, cultural, time-related, resource-related, and energy-related constraints, both inside the project team and in the relationship with other stakeholders, internal and external to our company. It was a matter of doing simple things that turned out to be very useful, sometimes decisive for the success of the projects.

For instance, promote frequent informal meetings among the components of the project team, asking different people to talk about the job to be done, everyone according to his or her personal point of view (which are the main difficulties and opportunities; what do they think they will earn from the project's experience; which similar experiences will they refer to in order to make it, etc.).

- Or, as a meeting's follow-up, to ask more than one person to write the meeting minutes, using his or her own language (try it: each minute might seem to refer to a different meeting).
- Or to keep a "road-journal," more or less private, to write our history of the project (using details, sensations, uncertainties, perplexities, incoherencies, etc.). And read it, when everything is over, together with other documents and bits and pieces of the official and of the unofficial history. And wonder how much of what happened we could have foreseen and whether the topics important for the success (or the failure) had indeed been the ones we had deemed as important at the beginning (or if everything was caused by a wing's beat on the other side of the planet).
- Or to pay more attention to the ways and means usually employed to communicate, avoiding typical overuse of words and tools: the terms (not so) out of fashion within professional jargon, which we still use for laziness; boring presentations; the Gantt charts that we use to deceive even ourselves; the name of the projects that have become acronyms and which should be meaningful, but cannot even be spelled; the e-mail messages where the subject is just written at random; half a dozen of persons in "CC," and nobody really accountable in "TO."
- Or to repeat things over and over, to whomever is concerned, in many different ways.
- Or to make sure that whoever is working on something new can update the others, when it is useful, using his or her own words and actively helping out.

We dealt with these and other things, taking care of our projects. And we did it not just for fun, but because of the need of managing "complex" situations. And so we got concrete and decisive results, above all, in the (continuous) definition of the project scope and in the strong involvement and contribution that we got from both the team and the other stakeholders.

We are not able to write a "guide for narration" yet. Maybe we would not even like to do it. And maybe another time we will talk about the technologies that can be used to narrate. For the time being, we can say that either the Moleskine, or the fragile memory of the PDA, or even the good old mailboxes, are enough. It is good to take pictures, to make movies, to design maps, and to write multisource minutes. Blogs, wikis, and project sites will do. We all know everything about the publish–post–comment–order–categorize–link–tag–search tools: that's even too much and it is worth investing time to find out what we really need day after day.

Indeed, we are reductive when we discuss the most suitable tools for archiving "everything" and for the detection of emerging themes. Depending on the implemented technologies, possible narrations and the opportunities of reanalyzing the history of the project completely change. On the use of powerful means supporting narration of the projects, some hints are presented in the following. Here we just want to eliminate two opposite objections we have come across: "I do not have the powerful means I need, and therefore …" and "I have enabled powerful means but I cannot get anything out of them. …" According to our opinion, both objections make little sense.

But, of course, technology is not enough. One of the most rewarding results of the time we have invested and the risks we have taken with our job is a certain security in relying on our ability of seizing the essential, at the right time, trusting upon the powerful means but above all upon practice, common sense, and intuition.

We believe that each project manager, compatibly with the context of the job, can discover his or her way to narrate or allow narration of the project, maybe rediscovering what he or she can do well and what he or she likes to do (it's a common belief that Homer composed little or even nothing: he just gathered the collective memory of the tales; indeed, he might not have even existed).

We would even like to say "rediscovering his or her own intelligence," which runs the risk of being flattened under stacks of standards and consolidated practices. Therefore there is no standard kit for the narrating project manager. In the following, we just try to focus on some discussion hints. They are just simple notes, divided into topics thus far, that make sense to us and which we would like to use as a basis for discussion to share with the reader.

NOTES ON PROJECT NARRATION

Stories

In every project, always and in any way, being aware or not, "stories" are told. All the stakeholders tell something about themselves, about their world, and their point of view on the project, using their own language, which the project manager must be able to understand and to speak. Stories that are told (also the ones less pertinent to the scope) can contain fundamental information for the management of the project; for this reason it is important that the project manager is eager to listen to them, to try to interpret them, and to keep track of them.

Some stories that are "not managed" (i.e., stories that the project manager does not listen to or does not understand or underestimates while they circulate inside the team, or the hosting organization, or the customer's organization) can seriously damage the project. Story-telling enables the formation and expression of personal points of view, the representation of difficult situations, the communication of things that cannot be told otherwise, and growth of the ability to seize and represent the details that matter.

The project forms its own language while it grows, a "family jargon." The birth of a shared language is fundamental for the success of the project. Exporting (telling) the story of the project from one world to another (i.e., project manager ↔ customers, team ↔ team, team ↔ performing organization, technicians ↔ humans) is a way to fetch, create, and spread the words needed to specify requirements, negotiate, decide, co-ordinate toward a scope under definition, get support, sell, and so on.

In this process, the project manager has a fundamental role: to enable mediation between the languages of the stakeholders (telling the project to each member in the proper language and helping each one to understand the others), and to collect valuable contributions to create a "project language" (including jokes, puns, nicknames, etc.). The project can be seen as a set of stories that are recounted "to believe": the sponsor, the project manager, from the start everyone co-operating in the project speaks of the things they would like to see happening in the future. They hope it can be like that. For this reason they "tell" and they "do as if" and also thanks to that they can really realize their hopes. This is, for us, one of the most rewarding facets of our profession.

How can you tell if someone is good at drawing, or if he can make progresses, especially for figurative design?
My teacher said that you have to look a lot at the model (the real thing) and a little to the drawing: "I understand if you are drawing well by the percentage of time you spend looking at the real thing with respect to the time you spend looking at your drawing. It is the real which drives the drawing and not the mental concept that you have and that you take into the drawing without observing reality any longer"

... What does it mean? Most of the time spent in communication must be devoted to listening, not talking. While planning, time must be spent to observe the reality, seizing the important things and not translating everything into a scheme which has little to do with reality.

FIGURE 10.7
The best storyteller/project manager.

Direct Speech

The project leaves traces (notes, intermediate versions of the documents, mail) that are not usually kept inside the archives of the "official history" of the project. Keeping these scraps (even in an unstructured way) can be useful to read the unfolding of the project in a more complete way, and to reread when the story is over, or when everything looks complete, but …

For instance, every e-mail can contain very important context elements that we often miss (or we just miss very important e-mails completely, because their object is not in line with the content and we consider them irrelevant at that time). Therefore to narrate the story of the project means above all "to not reject," but rather capture, store, tag, embed, keep track of even these sources: the "direct speech" of the project, reality prevailing over models and "recounting" itself better than the best storyteller/project manager (see Figure 10.7) could do.

Secret Notes

They always exist. Inside the drawers of the team members and of the other stakeholders. They can be extremely important. The project can even have a structured "shadow" documentation: the real plans referring to the ongoing activities, whereas the usual story based on a plan many months old has to be told to the customer.

The Project Manager Diary

A great thing. Even to reread what happened the other day when I was there. I was not sleeping, but I did not understand anything anyway. For the project manager it is a good thing to write "freely" his or her own vision of the risks of the project, even at the beginning, the first impressions. Even if it is just for himself or herself, and at the moment nobody is willing to listen. It is against the student syndrome (I know I will not make it, but I deny and forget it).

There's a Time to Tell and a Time to Plan

In general the things said during the development of a project become more and more determined. It is a good practice to use the most suitable tools for the time needed to allow this gradual forming and organizing of thought and knowledge (see Figure 10.8). The initial phases are particularly suitable to the use of a communication model free from predefined models. But it can also happen that the customer wants the project manager to come to the first meeting with a Gantt chart where everything has already been taken into account. In this case it is necessary to learn to manage the anxiety, to iterate, and to find the time and the way to explore further.

The uncertainty of the future also can be explained through the few words initially used in the WBS to define the activities that we plan to execute. In general, it is important to be aware of the narration potential of the words used in the traditional tools:

- The name of the project could be particularly evocative (considering that many stakeholders will just know the name of the project, and will try to understand something from it).
- The minutes of a meeting where a cheerful and co-operative climate (or a completely different one) occurred, can represent that climate.

A page of test is a very "complex" tool. It is suitable to represent details and feeble signs which should not be overlooked when it is necessary to form the initial impression...
Furthermore a white sheet to write on can help to avoid false preconceived assumptions: it is useful to re-consider the WBS from previous projects and to study the lessons learned, but it is also extremely useful to remember that every project is, by definition, unique.

FIGURE 10.8
Gradual forming and organizing of thought and knowledge.

- The meeting minutes can convey diverging points of view in a constructive way, in order to keep track and preserve this diversity as an advantage in project documentation in light of future potential evolutions of the scope/stakeholder.

Powerful Means

The tools made available by the internal company portals of semi-new generation allow more than what is implemented in most cases. They usually offer functionalities that allow an easy delegation of publishing authorizations, a flexible and free structuring of content and the possibility of detecting emerging topics.

Many people can co-operate in the documentation of the project's history, within dedicated areas and offering contributions according to their own competencies. And if the publisher of content in a project site really knows what it is all about, he or she will also be able to choose the right words for a blog, to define an effective tag, or to create a new link in a wiki. If the team is able to co-operate in this way, the use of tags, classifications, and links can represent a way of browsing that validates what we have said about the importance of the weak links, of the different points of view on the same topic, and so on.

The development of project sites within a company intranet on the basis of the emerging models can face several difficulties such as:

- A company (that may have invested a lot in hardware and licenses) expects guaranteed results from the intranet project.
- The fear of chaos leads to meting out carefully the authorizations to publish, and prefers organizations with rigid and hierarchical content.
- It cannot be taken for granted that every user is skilled enough not to consider the iteration with a site as a further obstacle to the sharing of the results of his or her job (which is a problem in itself).
- The rollout of this kind of activity calls for a great conviction of the project sponsor, for adequate and continuous technical support, for project managers who are really willing to try it in their projects.
- There are also minimum technical requirements to take into account: the cost of publication of the documents must be on the same order of magnitude as archiving on our own hard disk, the system must be highly reliable and always available; remote access capability is extremely useful, and in some cases mandatory.

Taking for granted a series of necessary conditions, we just highlight two important things that helped us in our everyday activities. First of all, structured and unstructured content models can co-exist and interact: the normalized documentation space and the space of chaos. Secondly, a proven path to help the start of these initiatives is to begin a Web 2.0 project in parallel to support project management as an adventure to recount. Discussing online about "how project sites should be designed" and "what we are doing to decide what to do," co-operating and communicating with the tools we want to use to build project sites is a great help to discovering their potential and to demystifying the limits and difficulties of using it (or finding out if they are real).

11

Risk and Complexity

Roberto Villa

CONTENTS

I have also thought of a model city from which I deduce all the others, Marco answered—It is a city made only of exceptions, exclusions, incongruities, contradictions. If such a city is the most improbable, by reducing the number of elements, we increase the probability that the city really exists.

The Invisible Cities by **Italo Calvino**

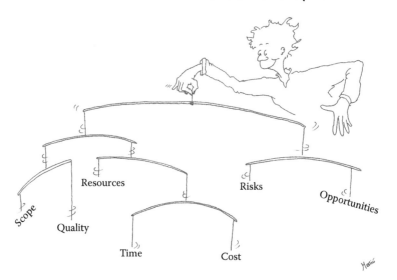

INTRODUCTION

Talking about risk in a project management context means hitting a nerve. Never before today has so much emphasis been placed on the fact that projects should be managed, yet things are far from well. Just take a look, among others, at the last Standish Group [1] research to see how the percentage of successful projects has not risen, indeed, in the last two years has even seen a slight contraction and still never fails to exceed a third of the projects analyzed.

Therefore risk management seems to be a panacea with respect to these results: because forecasts can so easily be missed, let's integrate them with the analysis of what could go wrong. But at the end we get to estimates that are unacceptable from every point of view—above all the stakeholders' one—so that we have to adjust it here and there to get to something tolerable, but no longer meaningful. Every project has, by definition almost, a certain degree of uncertainty of course: no matter how good models and lessons learned are applied, we still deal with real situations, which are subject to many variables whose behavior cannot be predicted, and sometimes they are not even identified! Indeed the project develops in a reality full of risks, some of which can be foreseen, and whose consequences can be inferred, whereas some others cannot (the so-called unknown unknowns).

More and more often we have to deal with vague, unstable, and even unstructured situations, without reassuring cause–effect relationships and characterized by weak links. The risk factor is overwhelming, or even innate, and the true word to describe the situation is no longer risk, but uncertainty. "A lot of good practices of project management can be thought as an effective management of uncertainty" [2, p. 3]. In fact we have uncertain estimates, and are also uncertain as to how to accomplish them. Sometimes we are uncertain about the real goals and how to weigh the priorities. We are always uncertain about the relationship between the components of the project. It is important to consider the uncertainty (understood as simply "absence of certainty") of every relevant thing in the project as a starting point for risk management.

Summing up, the five planning processes and the control process concerning risk suggested by the PMBoK®, as well as the various flavors of the two phases analysis–response and risk control suggested by other standards, are undoubtedly useful, but unfortunately concentrated on events, conditions, and circumstances, thus applicable only to some "lucky" situations, and to just a small part of those worlds of possibilities that unfold during the development of a project. Events are sparks that might indeed cause an explosion, but it is short-sighted to concentrate on the chance of having a spark without considering the degree of inflammability of the situation, and the general "health" conditions of the system, which are ultimately the causes for a specific fact to be dangerous or not. Furthermore, if we observe the project in the light of complexity, as we are trying to do in this book, it might even seem strange to extrapolate the concept of risk.

We can start from the "standard" definition of risk (we can use the one provided by the PMBoK as a reference), that is, "a possible event or condition which, if occurs, has an impact, either positive or negative, on the quality, the value or the schedule of a project" [5, p. 275]. If we apply it literally, it turns out that we have not described a risk, but rather the condition of a complex project. All our management skills should be addressed to the control of events and conditions that issue during the lifetime of the project, within a plan which is full of hopes and objectives, but also continuously influenced by reality as it unfolds, in a constant balance between stiffness and adaptability, linearity and redundancies, and aim for the objective and, above all, for a result.

BUT THINGS HAPPEN, THEN

In order to reach the essence of things, consider the project environment not only as a system, but, more completely, as an ecosystem, a system substantially in balance, whose various entities exist or transform, co-exist and link and disappear without generating significant perturbation to the whole. Consider it also as a context delimited by fuzzy boundaries (*um-welt*: what is all around) inside which a trembling reality unfolds in several forms relentlessly. In an ecosystem you can live, die, win, lose, eat, be eaten, build, and pull down. The whole does not substantially feel this turmoil, but uses it to get to a sort of quiescent state, a precarious, but stable, balance: on the edge of chaos, using an expression shared by all the Complexnauts, a homeostasis to use a more scientific term.

Phenomena of slow decay or sudden perturbations can occur as well, which can explode in sudden changes, leading the system to another point of relative balance. The situation can be exemplified by mountain brush, apparently amorphous to the eyes of a careless passer-by, but indeed with a deeper look full of both life and lifeless things: rocks, stones, water, ferns, roots, ants, insects, and the like. A world swarming about, sufficient to feed the imagination of a fantasy novelist, but which is not normally noted because of a substantial stability of the system, due to the overall balance governing it.

Of course, perturbations can occur and disturb this balance, slow or creepy, like the threat of a bacterium or the progressive impoverishment of humus, or sudden and violent, like a goat grazing moss, a hunter's foot, a landslide, whose effects, more or less evident, change the situation. In any case, the system tends to a new balance, either to the old or to a brand new situation, which is substantially stable. In the same way, our project can exhibit a brush of moods, jealousies, experiences, enthusiasm, skills, and so on that can positively influence its unfolding, even leading to success or failure.

Another significant example is the concept of harmony in music, meant as an expansion of the leading melody. Harmony comes from chords, often related by determined ratios, and it deeply influences the tone of the track. In classical Western music, main chords are consonant; all the dissonances are considered as a "passage": moments of tension that prepare the so-called perfect and decisive chords. We can therefore understand that our projects exhibit a virtuous balance, seen at a global level, which sometimes runs across a "wrong note," either in an evident or in a creepy way, which moves the project away from balance.

Assuming that the aim of the project manager is to control both the turmoil and the wrong notes, risk management therefore recognizes and enforces a status of homeostasis, that is, an overall balance, addressing above all the area of uncertainty. On the other hand, it should be able to recognize and collect perturbations in order to drive in the valleys of the edge of chaos. Accepting, if not welcoming, the uncertainty, calls for a complete change of attitude in whoever deals with projects, who is usually trained to eliminate or at least tolerate it; nevertheless, the advantages of facing uncertainty fairly are evident, inasmuch as, willy-nilly, it exists, and it cannot be neglected.

There is strong resistance, even from a social standpoint, to say "I don't know" and act accordingly, even if it would have been better many times to say so. How much time, and how much effort is spent by people and organizations to create false and partial certainties just to give the impression of control, which can generate illusions that are often devastating, or at least ineffective. Error is a way to get near the target: but while having an experience is unavoidable, learning from experience is not.

Convincing people, especially ourselves, that saying "I don't know" is possible, opens the doors to elaboration, to comparison, to knowledge, and to exploration: it allows us to distinguish between what is stable and our premises, to have a clear idea of the grounding of information. It facilitates discussion and knowledge-sharing and generates, even if it seems paradoxical, a climate of trust. Furthermore, it keeps attention high and leaves room for fantasy, that is, for innovation. Even if "the man is really eager to know what is certain" [3, p. 22], to keep a humble attitude toward knowledge and reality, and accept the right amount of uncertainty, it creates in ourselves and in the environment a substrate that favors the balance of the system and helps us in detecting and facing risks [4, p. 56]. The same attitude of openness should be kept with respect to perturbations to system balance.

A reductive vision of project success is the one that describes uncertainty and risk in terms of a threat to the success of the project. It is important to recall the importance of the concept of positive risk, that is, of opportunity. Good and bad fate has, a priori, the same probability, and any change is neither positive nor negative per se with respect to the objectives of the project, inasmuch as the change in situation can either facilitate or be an obstacle. These two facets co-exist in every situation of uncertainty, seldom in an independent manner, as two sides of the same coin, both threat and opportunity, and both of them must be managed. If these situations can be governed properly, the chance of getting to the target (maybe even moving it a bit) increases.

If we just concentrate on the probable event and its severity, it will be difficult to understand which opportunities can be generated, because we would only see negative consequences. And we would not be able to influence the evolution. It is necessary to see the situation from a broader point of view, embracing a more complete context, so that movement and change can be seen in a systemic way, and consequences can be understood and exploited.

We're often asked to make a forecast of the future as if we had a crystal ball, but we don't have it! We can just have a vague idea of the future or, even worse, just an illusion of it. Indeed, it is possible to make a forecast for a portion, an outcome of possible evolutions of a given situation, but the precise path is determined only by the context and by fate. In terms of risk, we can identify the range of possibilities as a form of compliance and work within this range using the approach of uncertainty.

Risk management in a complex context means then to practice and get trained for the responsibility of coping with emergences, that is, situations which arise, and also to educate with the aim of achieving the capability to cope with and respond to the situations of the project within the detected field of possibilities. This necessitates creating a culture and an attitude that encourages the uniqueness of each situation, rather than regularity and similitude, through common sense. This means, when on the water, adopting the behavior of a yachtsman, accepting the current, the sudden gusts, and the irregularity of the waves to exploit them as a thrust to get to the result.

The objectives of risk management are therefore the evaluation of the state of health of the system, to favor or to contrast the conservation of the overall balance, and the understanding and detection of the possible spaces of change in the system's balance to get ready and accept the change, thus improving the situation. All these topics make me think that risk management is an attitude to foster, not only a discipline to learn. Eric Hoffer said: "In moments of change (project) the earth will be of those who learn, while those who know will be well equipped to live in a world which does not exist any longer" [6].

From all that has been said, it brings to mind a toolbox which is life itself or, in other words, the person in his entirety, made of knowledge, experience, curiosity, wisdom, intuition, and emotions. I can try to provide some suggestions, from the many possible ideas, as an example that is far from being exhaustive, with the aim of reinforcing and underlining again the need of building a skill that is able to become an attitude in ourselves and in the team. These are tools that should not be used, but rather interpreted.

Method: Listen, Think, Contemplate, and Decide

If we want to build an attitude, we cannot start from a specific method: Figure 11.1 highlights some hints to help us get the adequate spaces and lead to an effective result. On many occasions, it is better to stay in wait mode allowing the situation to develop. It is worth noting that these are all active attitudes, which require our careful attention, except for the third moment. To listen is not to hear; it calls for our attention, but it does not mean to judge: it is a smart collection of what exists, which then has to be elaborated by thinking, trying to pull it out of context and see it under different perspectives: it is the moment for rhapsody, for narration; the use of metaphor is a good technique, because it clears the scenario of prejudices and influences and lets us see the situation under different lights, forcing us to have a broader look and enabling the use of fantasy.

The third moment is the one where we leave space for emotions and sensations: most of us make the most important decisions of our lives using

FIGURE 11.1
Moments and hints to face uncertainty.

our gut feeling (falling in love, having children). Why shouldn't we do the same with our job? It requires a careful listening to our own interiority, allowing the emergence of what comes out. At the end, all considered, we draw our conclusions, accepting the fact that we are not perfect, and we act.

Homeostatic Radar

Literally, homeostasis is the ability of an organism to maintain some values unaltered despite the changes in the surrounding environment. What these values and alterations are depends on the specific context. Therefore we can perform an analysis to uncover the most important values and plot them on a radar chart, qualitatively evaluating both their sensitivity and their touchiness to the context. In this way, an overall picture of the sensitive spots in our project can be depicted; some of these spots are indeed raw nerves, areas on which to focus our attention.

From the project charter, or even better from its manifesto, its peculiar and characteristic values can be found: on top of these values, we can add the ones detected by the sensitivity to the risk. Each of these values is represented with a fishbone diagram (Figure 11.2) in order to find both positive and negative influence values, those variables that can cause either its degeneration or its exaltation; depending on the corresponding weight, the overall touchiness index for the value can be found.

If the indices of the values are plotted on a radar (Kiviat) diagram, whose axes represent the maximum tendency to stability or the extreme touchiness to loss of stability, we can get at a glance an idea of the overall

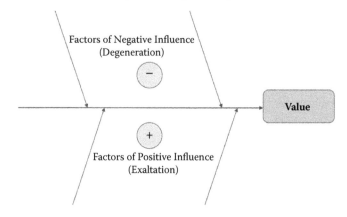

FIGURE 11.2
Touchiness index for value of the project.

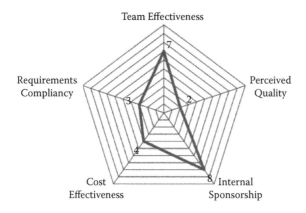

FIGURE 11.3
Example of homeostatic radar for a project.

stability of the project; moreover, it is possible to understand which values have the tendency to move away from the balance point, thus causing bad surprises (see Figure 11.3).

Consider, for example, the development of a website for a small company: possible values could be the budget and compliance to requirements. An informal internal procedure, the presence of more than one owner, and the use of an external consultant are all environmental factors which in that context could significantly alter our project: a look and feel the partners do not agree upon, overlapping roles between the internally responsible consultant and the external one, and poorly defined strategies. These are all elements that can turn a small project into a failure.

On the other hand, the presence of a smart administrative director can compensate for these disturbances if he is able to realign both partners and consultant as soon as costs go off the track. A touchiness index for the value (adherence to budget) of, say, 4 (on a scale of 1 to 10), allows us to consider that it will mostly self-govern and we can therefore pay little attention to it. The availability of a graphical map of these situations, our radar, is a relatively easy way to guess which area should be taken care of most.

Embrace Uncertainty

To manage uncertainty does not mean to manage identified threats and opportunities, but to find out all the possible sources for uncertainty within a context in order to explore and understand their origins; a prejudice-free approach about what is desirable and what is not is simply a must.

In some cases, it is advisable to keep uncertainty alive in order to stimulate creativity, thus having more options and degrees of freedom: decrease knowledge (the "hiding hand" mentioned by Hirschman) to increase possibilities. An interesting example can be found in the teaching trick used by Reverend Abbott in 1884 in Flatland, where a bidimensional world is described in detail in order to prove the existence of the fourth dimension: Flatlanders cannot conceive a tridimensional world, apart from some rare moments, as a manifestation of time. The author highlights in this way the same perplexities that we have when moving from the third to the fourth dimension.

Several approaches exist: the following has been suggested by Clampitt and DeKoch [4, p. 7] (see Table 11.1). Always keep in mind that governing and managing a movement in small steps is less difficult than either starting or stopping it.

TABLE 11.1

Summary of an Approach for Uncertainty Management

Steps to follow	Number	Action
	0	Accept uncertainty
	1	Define investigation domain
	2	Determine uncertainty level
	3	Clarify sources of uncertainty
	4	Define the approach
	5	Act accordingly
Levels to consider	**Degree of Uncertainty Levels**	**Methods of Uncertainty Levels**
	Laws	Authority
	Principles	Experience
	Rules	Instinct
	Premonitions	Reasoning
	Intuitions	Verification
	Unknowns	
Approaches to adopt	**Source**	**Attitude**
	Complete ignorance	Deepening, hypothesis
	Conscious ignorance	Investigation on specific topics, Plausible hypothesis, experimentations, explorations
	Fate	Overall picture, trial
	Complexity	Approximations, simulations, overall picture, indicators, research of fundamental schemes

The Wisdom of a Glance

It takes a peculiar style to be able to seize, among blurred contours, the distinctiveness of a suggestion, a specific detail that delivers the sense of a world among all the possible ones. Let's think about the detail of a face or of a hand in a picture representing a crowd or, more mundane, to a joke heard on a coffee break. Citing TS Eliot, every moment is the good one, where the eternal plot can reveal itself. Afterward we will be able to choose and say: "That was the time ..." [7, p. 149].

This glance, besides collecting unique things, should also be able to see the complete full picture. It should be able to see the whole framework to get the idea of what is happening. It should not lose the full picture and its perspective. Breadth and depth are not contradictory, but they sum up, reciprocally providing a complete meaning. One confirms or contradicts the other, developing its sense. But it is necessary to learn how to stop at the right time: when the whole can be understood, without chasing useless details, but at the same time without remaining too much on the surface. Wisdom and need help a lot to achieve this.

A simple example: I am on vacation in an apartment shared with other people and, on the first night, I realize that someone else has a toothbrush whose color and shape is the same as mine. In order to find elements to distinguish them, I take a deeper look and I see, for the first time, that mine has a vertical red stripe among the bristles, and the other has three horizontal stripes. For me, it's enough.

Ethic for Harmony, or a Shared Value System as a Basis for Compliance

The last hint is maybe the more important, inasmuch as it focuses on the creation of an effective (per se) environment for risk management, that is, to build and share a system of values inside the project which allows everyone to act and react naturally and with awareness when faced with the emergences generated by the project.

This is achieved by considering compliance as adequate to the environment, where each part is compatible with the others and with the context, so that it can easily adapt to the world as it is generated. It is worth devoting resources (time, cost) to this, rather than fighting against the risks, either presumed or present, through knowledge- and information-sharing, through direct and personal contact, and through listening and paying attention to other people.

It is necessary to give way to imagination in order to generate a repertoire of potential, of hypotheses, of what has not happened, and maybe will never happen, but which could have been, amplifying listening and intuition capabilities, getting a wide range of perspectives. It means to accept trust as a tool for a relationship, and to generate other trust through comparison and humility, which is, again, an attitude of listening.

The above-mentioned examples are just some of the ways to cope with risk management in a project context, and they go hand in hand with the traditional processes described by the PMBoK: plan risk management plan, identify risks, perform qualitative risk analysis, perform quantitative risk analysis, plan risk responses, and monitor and control risks. On a case-by-case basis, the wisdom of the project manager will choose the most appropriate tools: maybe more than one, applying redundancy, in order to achieve the best possible awareness of the situation and find the best way to follow.

REFERENCES

1. *Chaos Summary.* (2009). West Yarmouth, MA: The Standish Group International Inc.
2. Chapman, C. and Ward, C. (2004). *Project Risk Management,* 2nd edition. Hoboken, NJ: John Wiley and Sons.
3. Einstein, A. (1954). *Ideas and Opinions by Albert Einstein.* New York: Wings.
4. Clampitt, P. and DeKoch, R. (2001). *Embracing Uncertainty: The Essence of Leadership.* New York: M.E. Shape.
5. Project Management Institute (2008). *A Guide to the Project Management Body of Knowledge* (PMBoK® Guide), 4th edition. Newton Square, PA: Author. p. 275.
6. Hoffer, Eric. (1963/2006). *The Ordeal of Change.* New York: Harper and Row. Reprinted by Hopewell Publications.
7. Eliot, TS. (1934/1963). *Choruses from "The Rock."* In *Collected Poems 1909–1962.* New York: Harcourt Brace, 1934. London: Faber and Faber 1963.

12

The Value of Redundancy

Bruna Bergami

CONTENTS

Actually many of the blind men who tap their canes on Zirma's cobblestones are black; in every skyscraper there is someone going mad; all lunatics spend hours on cornices; there is no puma that some girl does not raise, as a whim. The city is redundant: it repeats itself so that something will stick in the mind.

The Invisible Cities by **Italo Calvino**

IS REDUNDANCY, LIKE ART, USELESS?

The term *redundancy*, in common acceptance, usually conveys a negative meaning, because it is associated with the concepts of superfluity and uselessness: what is useless is often perceived as a waste to be eliminated. Language, both spoken and written, tends to eliminate the superfluous.

Nobody uses a rhetorical language any longer. E-mail communications are so widespread and frequent that a concise and synthetic (nonredundant) style is used.

As project managers working in the computing and telecommunication fields know very well, systems and infrastructures are redundant, often employing duplication techniques in order to guarantee functionality even in emergency situations. Redundancy is carried out after the analysis of the risks of halt and crash of the systems according to the level of services that have to be guaranteed. Even in this case, redundancy is seen not as a plus, but as an excess, an inessential duplication that can normally be eliminated with no performance degradation, but is tolerated for emergency situations. It can be noted that the risk is meant as a potential negative effect on the system and, after its estimation, strategies to govern it are developed, starting from the assumption that the context, or better, the change of context, can be predicted.

During their careers, it is certain that many project managers have come across the process of modelization and normalization of relational databases, to eliminate data redundancy which occurs every time repeated data are stored unnecessarily. Normalization of databases has certainly answered to the need of using fewer memory resources, in a moment when memory resources were scarce, and have also optimized access. But later on organizations felt the need to use the analytical information contained in the different databases in order to increase their competitive advantage, reaggregating them in a different way and exploiting a business intelligence strategy that, by means of data warehouses and data marts, has led to data redundancy.

Furthermore, the demand for data used by business intelligence is growing more and more, so that the quantity of data maintained by companies for analytical purposes is growing every year, creating a vast array of distributed archives within the different organizations. Without any doubt, database normalization processes are mandatory, but optimization logic has not been able to answer to a new and unpredictable need, that of business intelligence. In a logic of continuous optimization, where everything is aimed at maximum efficiency and effectiveness, redundancy is a waste: wastes, inefficiencies, and deviations from the predefined model must be eliminated, or at least controlled.

In his masterpiece, *The Picture of Dorian Gray*, Oscar Wilde states: "The only excuse for making a useless thing is that one admires it intensely. All art is quite useless" [1]. Is redundancy, like art, useless?

━━━━━━━━━━

WHY REDUNDANCY

We live in a complex world, with fewer and fewer certainties, where the only thing we can be sure of seems to be change. The project manager is measured on his ability to plan, execute, and control the project aiming at both optimization and efficiency, often involved in a multiproject context, suffering from a constant scarcity of resources, and accompanied by a decreasing level of predictability. Not all the projects are complex, but they all belong to a wider complexity scenario that bears with it both new emergences and changes. Traditional project management tools have been developed in order to plan scope, schedule, and costs of a project, as well as to control them throughout its life cycle, assuming a linear approach able to forecast and control everything, risks included.

Both uncertainty and risk are dealt with in some phases of the project. According to the PMBoK®: "risk and uncertainty [...] are greatest at the start of the project. These factors decrease over the life of the project." [2] Uncertainty is mentioned elsewhere in the PMBoK (p. 158), in a section entitled "What-If Scenario Analysis": "This is an analysis of the question: What if the situation represented by scenario 'X' happens?" A schedule network analysis is performed using the schedule to compute the different scenarios, such as delaying a major component delivery, extending specific engineering durations, or introducing external factors, such as a strike or a change in the permission process. The outcome of the what-if scenario analysis can be used to assess the feasibility of the project schedule under adverse conditions, and in preparing contingency and response plans to overcome or mitigate the impact of unexpected situations.

The concept of contingency is similar to redundancy: it is not mandatory for project execution, but it can be used as a "reserve" in order to eliminate potential risks, which either are known or can be assumed looking at current knowledge and past history.

Duration estimates may include contingency reserves, (sometimes referred to as time reserves or buffers) into the overall project schedule to account for schedule uncertainty. The contingency reserve may be a percentage of the estimated activity duration, a fixed number of work periods, or may be developed by using quantitative analysis methods. As more precise information about the project becomes available, the contingency reserve may be used, reduced, or eliminated. Contingency should be clearly identified in schedule documentation.

Traditional project management tools rely on the fact that even unforeseen events can be controlled: this approach can be adopted whenever a substantially stable (although complicated) project environment exists. When the context either becomes unpredictable or changes too quickly, such tools might not be appropriate to react quickly to overcome the crisis. An excessive optimization approach leads to the loss of innovation capability. How many times is an idea or solution not pursued just because its impact cannot be quantified? And how many ideas or solutions do not emerge because they are trapped inside a project framework? Such ideas could turn out to be essential for a change in scenario.

At their highest levels, many organizations have understood the content of both creativity and unmanageability of complexity. A complex system is more capable of tolerating perturbations without losing its balance than a complicated one. Resilience is the feature that makes it more flexible and allows it to better adapt to uncertainty situations. Component redundancy is the feature of a complex system that allows this flexibility.

Upon observing complex systems such as biological ones, as seen through system dynamics, some considerations on redundancy can be made. Both chance and need are elements present in current scientific models of biological evolution. Equilibrium theory, which states that evolution occurs by means of slow and imperceptible steps, is disproved by catastrophe theory, which states that global catastrophes create discontinuity. In such moments of crisis, survival depends on the capability that living beings have of adapting to new conditions. Such adaptation often relies on the recovery of some unused apparatus.

A well-known example (see Figure 12.1), which highlights the capability of biological systems to reuse and recover unused apparatus is the so-called Panda Principle, stated by Stephen Jay Gould.

> Panda are herbivorous descendants of carnivorous bears. Their real anatomical thumbs had been irreversibly devoted to limited movements which are common for all carnivorous mammals. When they adapted to a bamboo-based diet, more sophisticated manipulation ability was needed, but there was no way to redesign the thumb. The sub-solution implied the use of an enlarged radial sesamoid bone which is present in the *carpus* (wrist), the so-called "panda's false thumb." The sesamoid thumb is just an approximation, far from being optimal, but it works. [3, p. 59]

The decoding of our genetic code has shown that redundant information is embedded, apparently useless, but which probably allowed us to

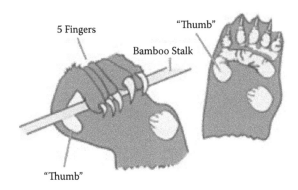

FIGURE 12.1
Adaptive evolution of the panda's thumb.

adapt in case of crisis. Therefore for living beings redundancy or duplication of some apparatus contributes to both robustness and flexibility of the organisms. Even if several components are lost, the system is still able to survive. Whenever the impact is so severe that it overcomes the maximum degree of flexibility, the system evolves to another balance. Redundancy is therefore a resource embedded in the system, essential in the case of unpredictable situations.

The fact that the panda devised a sixth finger out of an apparently useless bone of the wrist provides valuable information about evolutionary processes that are the foundation of biological diversity, and also provides a fresh perspective on several systems.

CREATE REDUNDANCY INSIDE PROJECTS

Taking biological systems as an example, it is easy to think that even complex projects, or even noncomplex projects within a scenario as complex as the present one, can benefit from redundancy. But which type of redundancy and which tools should be used? How does suboptimal redundancy compare with respect to traditional project management tools, which tend to model and optimize project components? How can redundancy be added within a resource scarcity framework?

We believe that embedding redundancy in a project does not mean to turn the traditional approach upside down completely, even if a substantial change of the point of view is needed. It is important to accept the fact that not all the facets of the project can be managed using traditional schemes, and that even something outside traditional schemes can be useful. The answer lies first of all in a different attitude and project vision, according to a systemic approach.

Even with "stable," noncomplex, simple projects, project managers experience every day the challenge of keeping the project on track, so that it can evolve according to the forecast. The study of biological systems has offered some general principles that can be applied to the project system as well, in order to keep it in an equilibrium state between compliance to standards and diversity, on the verge of chaos:

- Inject redundancy, trying to keep the right balance with the stiffness of traditional rules
- Inject maladjustment, exploiting it as a learning opportunity

But how is it possible to invest in redundancy while cutting costs? It looks like a contradiction if we do not change the perspective used to ask the question. We should try to simplify the less-valuable activities, which are resource-consuming, without being obsessed by reduction at any rate. At the same time it is advisable to invest in redundancy, that is, to accumulate resources which can be used to react to changes, quickly creating potential alternatives. Our proposal is to focus on knowledge improvement. To create redundancy means to increase interconnections, information, and experiences, to deal with different points of view, and to get used to seeking new ways.

In order to create redundancy, it is not sufficient either to use project management tools differently or add further reserves increasing contingency, which could lead to additional costs without bringing a significant benefit in terms of response to "unpredictable" events. There is neither a unique solution nor a unique tool, and our journey in complexity has just started. Nevertheless we can already provide project managers with suggestions and a proposal to "inject redundancy" into some portions of the project:

- Information
- Interactions
- Experiences
- Resources
- Approaches

Redundancy of Information

We can keep all information. Each project has its own way of filing information, "official" and "unofficial." How information is organized and archived depends on the organization, on the project culture, on the team, and on the stakeholders. Official information is usually archived in a structured and univocal way. Other information is considered useless, incomplete, or excessive, and is therefore eliminated or lost along the way. But even these "redundancies" are a valuable source of knowledge, and the narration of what is going on and of how we live it. These data shall be preserved, without giving them a precise structure, but simply tracking and tagging them to allow a proper search later on.

Even without considering the effort, undoubtedly considerable, required to structure data, it would be useless, and limiting, to store and organize information, building a scheme structured a priori like a database, or an approach based on a structured ontology. Even if the latter allows a greater usability and wealth of contents with respect to the use of simple queries, and to the aridity of the information of the scheme of a relational database, it assumes being able to describe the domains of interest that are unknown unless we make an initial assumption on the use of the contents.

We do not know when information becomes useful, but we know that we have it. Summing up, we suggest following the approach that does not erase information and traces without bonding, filing "with lightness." With respect to the pioneers of project management, today we can benefit from information technology that helps store any kind of document using simple and cheap tools.

Labeling or tagging a document by associating a few simple keys with it allows our finding it when needed. A search engine is enough to find it. One of the most widely used is Google-Desktop, which can be used to find a file at any time both in local and distributed archives, but many others can be found on the Web. Then for whoever wants to use paper notes, the possibility of digitalizing a document for electronic filing is a way to preserve not only the content, but also the form of the document, in order to follow and share a path, reusing it in other moments and other forms.

Redundancy of information also means considering those contents not strictly related to the context of the project. To move around in the world of complexity means to be able to search for and grasp new opportunities. Different experiences and different specialties that are apparently useless let us seize new visions and solutions to the problems. Each source can be useful; it is sufficient to practice and listen and perceive the signals coming from the world around us.

Even requirements are important information, not only because they help us define what we are going to implement, but also because they represent the needs, the points of view, and the features of the various stakeholders (see Chapter 7, "Stakeholders' Worlds"). That's why a vast and abundant collection of requirements improves our comprehension of the project. As is always the case, only a portion of the requirements is implemented as is; many of them will change because the needs change and because of the subsequent iterations and mediations between the interests at stake. If requirements are limited starting from the assumption that we cannot do everything and that we cannot achieve the impossible and therefore it is useless to take it into account, we risk and limit the perspective of the project. Quite often requirements that look useless allow us to go beyond the limits and to have a wider view and a wider spectrum of the possible project solutions.

Furthermore the collection of requirements should not be considered as the mere formalization of the requests and needs expressed by the stakeholders. Requirements are not only the ones embedded in formal documents, but those integrated with all the informal information collected in the project. It could just seem a waste of time, but how often we waste project time creating useless documents that nobody will ever read or attending long and chaotic meetings to decide things that could be sorted out in few minutes? Maybe it is not a waste to start using our time in a different and more productive way.

Redundancy of Interactions

A complex project can be seen as an open system featuring many internal and external connections among which an informal network of relationships exists (see Chapter 7, "Stakeholders' Worlds") that constitutes one of its intrinsic values.

In moments of crisis, a centralized system tends to collapse and therefore the relationship model must be distributed to guarantee flexibility and to respond to the increased complexity of the system (organization on the verge of chaos) with the right balance between formal and informal systems where a redundancy of relationships, information, and knowledge exists. Redundancy of relationships creates shared knowledge and continuous learning. The project becomes a learning organization, not only based on historical data and past trends, but also on continuous learning which leads to the exploration and generation of new perspectives.

Redundancy of interactions favors the valorization of the tacit knowledge inside projects. Depending on the degree of accessibility we can divide knowledge into tacit and explicit. Briefly, we can say that tacit knowledge is an uncoded knowledge, nonlinear, not managed through structured but informal communication flows, specific to an individual, and in relationship with his intuitions, feelings, and ability to relate to the context. Tacit knowledge is hardly accessible and difficult to describe. Citing E. Polanyi, precursor in this field, we can state: "We know more than we can say" [4, p. 42].

Explicit knowledge is formal, systematic, scientific, organized, and therefore easily understandable and accessible. Tacit knowledge includes cognitive elements, that is, mind models, schemes, paradigms, beliefs that help people to perceive the context, and technical elements concerning the know-how and concrete abilities. Tacit knowledge is a personal legacy of any individual, a hidden potential very hard to exploit.

According to the studies by Nonaka and Takeuchi [5], it is possible to maintain that an interaction exists between tacit and explicit knowledge, inasmuch as these two types of knowledge are complementary dimensions that interact in any creative activity of the individuals. Such interaction can be defined as a knowledge conversion allowing the exteriorization of knowledge through complex forms of communication.

One of the mechanisms or knowledge enabler that allows this conversion is represented by the redundancy of interactions. The presence of a multitude of interaction mechanisms promotes and accelerates the spread of knowledge. For this purpose it is fundamental to create, in the project context, conditions favoring informal relationships between individuals, because these relationships eliminate the barriers that impede knowledge-sharing and facilitate the creation of new knowledge through discussions and comparisons. In order to operate in this sense, it is useful to create a project culture oriented to participation and sharing of common goals.

Using the so-called "agile methodologies" used in software development as a starting point, which are partially inspired by complexity theory, we can say that a loosening of formal control facilitates redundancy of relationship, because the groups self-organize, facilitating the exchange and growth of knowledge; on the other hand, redundancy of relationships makes a strict formal control superfluous. The project manager should be more a leader than a manager, facilitating rather than controlling.

Contrary to agile methodologies (specific for the optimization of software design), where informal relationships are facilitated within the project team, but not toward the external world, we can assume that relationships encompass all the stakeholders, with the aim of increasing global system knowledge improving the exchange of knowledge topics developed in different application domains. In other words, interactions between different domains rather than between specialists in the same domain, for instance software development, increase the chances of discovering new knowledge. These relational mechanisms are quite complex and might lead to a potential information overload. In order to avoid an "excessive" redundancy, simple behavioral mechanisms should be shared, facilitating regulation, and integrating uncoded tacit flows with the coded and structured flows defined by project organization. Under this assumption, even project management practices can be partially distributed. For example, instead of top-down centralized planning implemented by heavily structured project teams, we can think of a development of detail planning that is decentralized and local and involves a greater number of people, such as the network of suppliers.

Even in this case redundancy of relationship can appear superfluous but, in a global view of the project system, wasted time is recovered. A greater involvement and therefore a greater knowledge and information sharing ensures, even in stable contexts, yielding better co-operation and limiting the misunderstandings that are often a hurdle to the achievement of project goals.

It should be added that physical proximity facilitates relationships, above all informal ones, but when being near is either impossible or too expensive many different distributed ICT infrastructures can be used to create virtual networks that, even with limitations, can facilitate the development of informal relationships and promote co-operation as an organizational practice. Concerning this topic, an extreme example of effectiveness of redundancy of relationship is represented by the way the Linux operating system has been developed: that is, a "bazaar"-based self-organization [6, p. 29], in contrast to the cathedral model which is common to every proprietary software development project, characterized by centralized planning and heavily structured project teams. Linux has been developed in a decentralized manner, based on the involvement of a high number of participants, a complete openness of the project, and continuous feedback, which have allowed the emergence through the network of distributed creativity and a collective intelligence.

Redundancy of Competences

To inject redundancy in projects also means to insert disruption (i.e., different skills), and to exploit the resulting maladjustment as a learning opportunity. In other words, diversity of experiences and skills facilitates innovation, contrary to the homologization of cultures, skills, and styles. In times of crisis, culture is a survival tool, but generalization has to be preferred to specialization.

Our culture is mainly based on specialization, which means that inside a project we risk cultural homologation, which leads to a "weakening." The advice is therefore to favor diversity between individuals and strengthen interactions between different specialists. This can appear as a way to impoverish communication, but confrontation between different information can lead to a more correct relationship with the plurality of cultures that compose it.

In order to carry out redundancy of competences, the project manager should facilitate role flexibility inside the project, that is, favor the creation of cross-functional teams highly focused on the vision and the objectives of the project, with less formalized assignments, where the assignment of tasks derives more from the self-organization of the team itself than from roles that have been imposed. This approach works very well in software projects, but it can be applied in different scenarios as well. In this way the development of skills of both the system and the individuals is facilitated, as well as a greater flexibility in resource planning.

Finally, talking of redundancy of resources, our project is one of the most appropriate examples. Our journey, what we have learned and the conclusions we came to, wouldn't have been possible without comparing all our different experiences, different visions, and contributions of experts coming from different disciplines, the research, and the re-elaboration of heterogeneous sources. After years of "specialization" where different disciplines including philosophy, biology, painting, and literature have been considered as personal hobbies and lateral knowledge, we started to consider them as an integral part of our everyday job. We have learned to revalue our legacy of "redundant" competencies, finding out that they are, or will be, often useful in our projects.

That's why even reading a novel can help us to work better on a project [7, p. 16]. Our ability to cope with new and difficult situations is developed not only thanks to assiduous study, or referring to past experiences. The uncertain present can be effectively lived only if our mind, working without constraints, in the way of emotion, brings to light here and now the knowledge that we need in this moment, knowledge that we may have accumulated by chance, with pleasure, for fun [7, p. 44].

Redundancy of Resources

An iterative approach to projects is the best defense against unpredictability. This statement is not new, and it fits very well in this context. Contrary to a linear approach, good for projects that are relatively easy, compact, and modular, an iterative mode is dynamic and facilitates adaptation to changes. In particular, a rolling wave planning technique is particularly useful where project scope is highly variable, when vast areas of the scope are either not well defined or unknown. In these cases a detailed and complete planning of the whole project, no matter how accurate, will exhibit a high degree of uncertainty in the best case, and will call for continuous changes with respect to the initial baseline.

Using the rolling wave technique [2, p. 120], the project unfolds according to a sequence of iterations. Long-term activities, in the initial phase of the whole project, are planned approximately, that is, at a relatively high level of the WBS. Detailed planning of the activities to be performed in the next and near iterations is completed as soon as the ongoing activity is completed. In this way, more realistic planning can be done, which is less subject to continuous changes, inasmuch as the focus is on what is presently predictable according to the requirements that emerge and evolve, as the project unfolds. Each iteration has an established pace, which is generally set according to the type and complexity of the project, and which in some cases is organization-specific.

Using this "plan as you go" approach, a rough estimate of the whole project (included resources needed) is done in the initial phase for each of the different horizons for its completion. The estimate is done top-down and it cannot reach a high level of detail, because it is based on approximate data. For a single iteration, the estimate is bottom-up and it details each single task. For each subsequent iteration a new estimate is repeated for the remaining iterations. The estimate has to take into account the remaining scope, which depends on how much has been released in previous iterations, on the variants and priorities of requirements, on the "learning" capability of the project team, evaluated with respect to previous steps. For the nearest iteration to be performed, the estimate is done with a higher level of detail and a lower "variance." Also the concept of

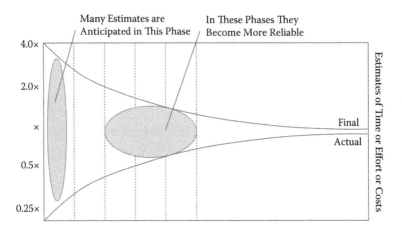

FIGURE 12.2
The evolution of project estimates over time.

baseline should be interpreted in a different way: we can talk of several baselines, one for each iteration, whereas a baseline for the whole project makes sense only when most of the estimated job has been carried out. As shown in the example (see Figure 12.2), the estimate becomes reliable only after a certain number of iterations.

From what we have said, we can draw some conclusions. Rolling wave planning requires the project manager to get used to operating with both accurate and imprecise data at the same time. The quality and the usability of the estimates for the whole project are proportional to the effort required to implement them (it is different to plan some hours or some days); on the other hand they always exhibit a considerable degree of uncertainty no matter how big the effort. Along with the base assumption, we are moving in an unstable context. The project manager is therefore in the paradox of formulating an estimate of the resources needed by the project as accurately as possible without having precise estimates. But an estimate should be done, otherwise, as stated by Dwight D. Eisenhower [8]: "In preparing for battle I have always found that plans are useless, but planning is indispensable."

In some cases, a time-boxing technique is used, keeping for each iteration the number of team resources fixed, and varying the number of iterations required to complete the project. Regardless of the technique used, we want to highlight the possibility of introducing in resource estimation the concept of redundancy which grants a greater autonomy in project execution. The rolling-wave technique, which is a way to adapt the project to the changes of the real world, calls for flexible resource planning as well, due to the greater degree of uncertainty of the estimates as well as to the higher number of intersections that occur between the phases of the life cycle of the project.

It means to account for a resource virtual pool to be used when needed, coherently with the phase or gate development approach used, and it often obliges us to keep all the resources available for the whole project, or at least for more than one phase.

That's why some organizations have resource pools shared among different projects, in order to ensure an easier reallocation of resources from one project to another, depending on the needs. It should be added that with the creation of competence redundancy, thanks to the flexibility of project roles, a resource redundancy is created as well, which allows an easier adaptation to changes of planning.

Redundancy of Approaches

In many public and private companies, the need for a more accurate control of the project life cycle has led to the implementation of common methodologies or reference models based on the standardization of processes and related documents (plans, specifications, etc.).

It should be said that the effect of such choices on the improvement of project performances has not always been as good as expected, even if we should recognize that the benefits introduced by the use of a "common language" are tangible and often relevant. Certainly the increasing uncertainty and variability of project scenarios during the last decade have clearly shown the limits of standardized approaches, reducing their beneficial effects as far as overall efficiency and effectiveness are concerned.

One of the answers to contrast this tendency has been the adoption, in the light of redundancy, of a methodological reference model, which gives us the opportunity of profiling project evolution according to the features and complexity of the project itself (the adoption of the so-called agile methodologies is an example of an alternate model used for some types of projects, mainly for software development).

With respect to this kind of choice, what we want to propose here as a form of redundancy of approaches is something more substantial and, to some extent, cultural. In order to clarify the genesis of our proposal, it is advisable to start from an assumption that the French sinologist François Jullien has described very well in one of his books [9, p. 12]. In short, he states that the foundation of the Western way of thinking and of classical science which allowed its extraordinary development is the idea that any strategy shall be based on the pre-emptive construction of a "model" of reality and on the subsequent implementation of such a model through a plan aimed at the objective to be reached. The battle a military strategist simulates on paper as in a sort of Risiko game, or the "mythical" provisional budget [10, p. 21] on which many managers have anxiously bet their careers, are just some examples of the migration from theory to practice through the sequence of "design and implementation."

Looking at the Chinese way of thinking, we come across something that is very far from our reference points, due to the fact that the history of China developed in complete autonomy at least until the sixteenth century, without those contaminations that have affected other Eastern countries such as India (to this end, we should remember that Sanskrit belongs to the Indo-European linguistic stock). The Chinese way of thinking approaches strategy in an opposite way with respect to Western cultures. There is no trace of a prebuilt model, therefore reality is simply observed as it happens for what it is with a unique aim: to evaluate the so-called "potential" of the situation, to identify the factors that best facilitate the achievement of the objective. Therefore the subsequent strategy is not carried on by means of a plan, but rather by actions aimed at progressively exploiting the favorable factors (also known as "bearing" factors) trying at the same time to turn unfavorable ones around. If we recall the previous example of military strategy and we compare it to the renowned texts on the art of war written by the Chinese generals Sun Tsu and Sun Bin, we can easily spot the peculiarities of an approach based on the sequence "evaluation and exploitation."

Many other words could be spent on the deep differences between the two approaches of "design and implementation" on one hand and "evaluation and exploitation." The application of the two strategies in a project context highlights such differences even more, as far as actions and behaviors are concerned (just consider the shift from a logic framework of "means–ends" to a logic framework of "conditions–consequences").

And this is the translation into practice of the principle of redundancy, meant as the ability to make the two opposites co-exist, that is, to be able to adopt one approach or the other depending on the features of the project or, even better, as a consequence of relevant changes in the project framework. Of course the project manager must be the first one to adopt such flexibility and which, in my opinion, should be one of the distinctive skills to be sought after in the selection process of professionals hired to manage projects characterized by a high degree of complexity and variable contexts.

But how can we effectively implement this ability to apply redundancy to the traditional approach combining it with a strategy inspired by binomial evaluation and exploitation? In previous chapters there are many hints to take inspiration from, and it is useful hereafter to recall the most significant ones:

- A first example is to focus on the "stakeholders' worlds," trying to understand the premises, to know the history and the role within the relationship network to which they belong (see Chapter 7, "Stakeholders' Worlds"). Which key stakeholder (first of all, the sponsor) is the best one to suggest the "bearing factors" that we can use to be trailed by, like a surfer on the waves, to the objective? A careful evaluation of the reality they live in is the best way to find out, beyond their needs and demands, their true expectations and dreams for which they would give us what they usually do not. We do not mean to imply that we have to abandon the references created by our model of reality (WBS) and the corresponding plan, but we need to understand that the identification of a favorable factor to exploit a project's advantage cannot be sacrificed to the so-called "triple constraint."
- Another example of strategy inspired by the typical Chinese way of thinking and, more generally, to the Eastern one, lies in the attention devoted to the "favorable time" (see Chapter 8, "The Propitious Time"), that is, the ability of interpreting the temporal dimension not only in a chronological way. Even in this case we do not ask the project manager to throw the Gantt away or to replace it with "sail at sight" tools, but rather to make it redundant with information more related to the time of circumstances or *kairós*, for instance, trying to highlight those windows of opportunity that could open up in some phases of the project.
- Last but not least, another example of the redundancy approach with respect to the traditional one is related to the set of tools proposed in the chapter discussing the "systemic" view of the WBS (see Chapter 6, "The Project beyond WBS"). In fact the emergences diagram and the space of domains are nothing but practical ways to evaluate the evolution of the project scope as it unfolds, to quickly understand the evolutionary trends in order to take advantage as far as the governance of the whole project is concerned.

Summing up, the redundancy of approaches defined as the ability of reading reality and acting upon it starting from different perspectives is by no means a way to introduce contingency in terms of time and cost, but indeed can represent a sort of balancing tool that can be used by the project manager to keep the equilibrium on the so-called "edge of chaos."

REFERENCES

1. Wilde, Oscar. Memorable Quotes and quotations from Oscar Wilde. http://www. memorable-quotes.com/ oscar+wilde,a368.html.
2. Project Management Institute (2008). *A Guide to the Project Management Body of Knowledge*—PMBoK® Guide. Newton Square, PA: Author, pp. 120, 158.
3. Gould, S. J. (2001). *Bully for Brontosaurus: Reflections in Natural History*. New York: Vintage, New Edition, p. 59.
4. Polanyi, M. (1974). *Personal Knowledge*. Chicago: The University of Chicago Press, p. 42.
5. Nonaka, Ikujiro and Takeuchi, Hirotaka. (1995). *The Knowledge-Creating Company: How Japanese Companies Create the Dynamics of Innovation*. New York: Oxford University Press.
6. Kuwabara, K. (2000). Linux: A Bazaar at the Edge of Chaos. *First Monday*, 5(3): 29.
7. Varanini, F. (2007). *Leggere per lavorare bene*. Venice: Marsilio, pp. 16, 44.
8. Eisenhower, Dwight D. Memorable Quotes and quotations from Dwight D. Eisenhower. http://www.memorable-quotes.com/dwight+d++eisenhower,a142.html.
9. Jullien, F. (2005). *Pensare l'efficacia in Cina e in Occidente*. Rome: Laterza, p. 12.
10. Varanini, F. (1994). *T'adoriam budget divino*. Milan: Sperling & Kupfer, p. 21.

13

An Ongoing Journey

Walter Ginevri

CONTENTS

"Journeys to relive your past?" was the Khan's question at this point, a question which could also have been formulated: "Journeys to recover your future?"

And Marco's answer was: "Elsewhere is a negative mirror. The traveller recognizes the little that is his, discovering the much he has not had and will never have."

The Invisible Cities by **Italo Calvino**

WHERE WE STARTED FROM

Even if I am in charge of writing the last chapter, I would like to clarify that my aim is not to "close" the topic (which would indeed disown one of the fundamental principles of complexity, i.e., try to learn), but rather to restate the need to open up project management to new developments: in fact, as stated by most of the authors of the first part of the book, project management remains a central managerial discipline.

In order to carry out this task I revisit the outcomes suggested by each Complexnaut in the light of my experience as a project manager, and try to summarize the relationships among the topics discussed in the previous chapters and the main subjects of the PMBoK® (knowledge areas, life cycle, processes, tools and techniques, etc.), that is, the most widespread reference model in terms of project's best practices.

Before starting, I would like to clarify the path that led to the emergence of the previously developed seven topics among many others. Despite all the limits of a retrospective analysis, I think that the hints which inspired the research team most have been the following:

- A short essay written by our mentor Francesco Varanini [1] which has been an introduction to the complexity topic, showing the intimate connections with many project issues and anticipating something about the journey we were about to begin. Reading that essay again at a later time, I can definitely say that the correspondence between the "expected reality" and the "actual reality" lived by the Complexnauts has been incredible, to say the least (I wish travel agency brochures were as truthful).
- Among the several books used as references for study by the team, those written by De Toni and Comello [2,3] allowed us to understand the basic principles of complexity theory, and therefore we have been able to apply them within project contexts.

Thanks to both the previous essay and many other readings recommended by our mentor, as well as the discussion with "experienced mariners" who have provided their witness to this book, the research team has been continuously stimulated: curiosity, anxiety, and surprise have always been common feelings.

In a totally unpredictable, and therefore perfectly in line with the try-to-learn approach, a huge number of questions have emerged, which have highlighted the limits of traditional best practices while setting off the search for further integration of said practices. Summing up, here are the "emerging questions."

- *Scope* (see Chapter 6, "The Project beyond WBS"): "How can we overcome the limits of a 'reductionist' vision of the project scope, whose management is heavily based on a logic of control?"
- *Stakeholder* (see Chapter 7, "Stakeholders' Worlds"): "How can we overcome the limits of a structured approach that is just developed through the identification of their needs?"

- *Time* (see Chapter 8, "The Propitious Time"): "How can we overcome the limits of a purely 'chronological' arrangement and sequence of project activities and actions?"
- *Leadership* (see Chapter 9, "Leadership and Complexity"): "How can we overcome the limits of an 'individualist' vision, based on a set of attitudes and skills implemented by the project manager?"
- *Lessons learned* (see Chapter 10, "Narrating to Believe"): "How can we overcome the limits of so-called storytelling, which is often characterized by prepackaged recipes (questionnaires, templates, rituals, etc.)?"
- *Risk* (see Chapter 11, "Risk and Complexity"): "How can we overcome the limits of approaches based on both probabilistic analysis and response strategies uniquely inspired by the idea of contingency?"
- *Resources* (see Chapter 12, "The Value of Redundancy"): "How can we overcome the limits of 'optimized' (and therefore not redundant) approaches that do not enable the ability of responding to changes?"

These questions might even be compressed into a sort of paradox, which would go like this: "Does it make sense to focus on recurring and repetitive events of a project, which at the same time has been characterized as something unique?"

Let's now highlight the relationships between the topics that we have investigated and fundamental principles of complexity, which are well described in the books by De Toni and Comello [2, pp. 81–223]: a correlation matrix (see Table 13.1) can be developed, which shows several links and implications worth considering. If there are still doubts about the similarity between complexity theory and the project domain, a "name game" can be played, using the following statements that the team has collected from several essays written by authoritative experts on complexity:

- C. Langton: "A *complex system* looks for the edge of chaos, i.e., a vital state which is both in dynamic equilibrium and never-ending change between order and disorder" [3, p. 110].
- J. Holland: "*Co-evolution* is a dance which develops through a subtle game of competition and cooperation, creation and reciprocal adaptation" [3, p. 211].
- H. R. Maturana and F. J. Varela: "*Knowledge* is not a representation of a world which exists on its own, but rather a continuous creation of a world" [3, p. 129].

TABLE 13.1

Relationships between Investigated Topics and Fundamental Principles of Complexity

Investigation Topics ⇨ ⬇ Complexity Principles	The Project beyond WBS	Stakeholders' Worlds	The Propitious Time	Leadership and Complexity	Narrating and Believing	Risk and Complexity	The Value of Redundancy
1. Self-organization: spontaneous emergence of new structures	☑		☑				☑
2. Edge of chaos: state of dynamic balance between order and disorder	☑		☑		☑		
3. Hologramatic principle: each thing is in the universe and the universe is in each thing	☑						☑
4. Impossibility of forecast: boundary state between predictability and unpredictability	☑	☑				☑	
5. Power of connections: everything is connected to everything else	☑				☑		☑
6. Circular causality: the effect has a feedback on its own cause			☑			☑	
7. Try-to-learn approach: trial and error is the only way to learn				☑	☑		

Now, if we try to substitute the words in italics with the word "project," we would realize that every statement perfectly fits the reality that any project manager faces every day.

WHERE WE ARE NOW

One year and a half after the kick-off of our project, we can say that we have traveled a fair amount of the journey, and our knowledge has grown considerably from a qualitative point of view, thanks to the high degree of multidisciplinary approaches that have characterized the research. At the same time, the breadth and depth of the topics which emerged have been so evident that they proved the aptness of the following statement made by a renowned expert [4, p. 113] in complexity: "Every increase in knowledge corresponds to an increase in ignorance."

Starting from these premises, let's now see where we are with respect to one of the main objectives of this research activity, the identification of approaches and tools that can be used to both integrate and enrich the framework settled by the PMBoK. A first element for comparison can be depicted by mapping the relationships existing between the knowledge areas of the PMBoK (e.g., project time management) and the seven topics on which the research team focused. Table 13.2, without claiming to map a precise intersection between the two domains, highlights how such topics have strong implications for the process with reference to the specific knowledge area (e.g., the relationship between the theme of narrating and communication management processes).

The last and maybe most interesting comparison between the outcomes of the research and the best practices of the PMBoK is the one related to project output that has been identified by the research team and described in some of the previous chapters. It is important to highlight the complementarity of such output, inasmuch as the research, since its inception,

TABLE 13.2

Relationship between the Knowledge Areas of the PMBoK and the Research Topics

Investigation Topics ⇨ / ⇩ PMBo K® Knowledge Areas	The Project beyond WBS	Stakeholders' Worlds	The Propitious Time	Leadership and Complexity	Narrating and Believing	Risk and Complexity	The Value of Redundancy
1. Project Integration Management	✓	☑		✓	✓		✓
2. Project Scope Management	☑	✓					✓
3. Project Time Management	✓		☑			✓	✓
4. Project Cost Management	✓					✓	✓
5. Project Quality Management	✓						☑
6. Project Human Resource Management		✓		☑			✓
7. Project Communications Management		✓		✓	☑	✓	✓
8. Project Risk Management	✓	✓	✓			☑	✓
9. Project Procurement Management		✓					✓

TABLE 13.3

Relationship between the Outputs of the PMBoK and the Topics Investigated by the Research

PMBoK: Process Groups and Main Output		Complexity: Complementary Outputs and Domains			
Initiating	*Project Charter*	Emerging Project Charter	Project Manifesto		Project Building
Planning	*Work Breakdown Structure*	Project Systemic Map		Project Narration	Project Driving
Executing	*Issue Log*	Emergences Diagram			
Monitor/ Control	*Performance Reports*	Space of Domains			
Closing	*Lessons Learned*	Project Narration			

has followed an evolutionary idea rather than establishing a discontinuity with respect to traditional approaches.

Under these assumptions, relationships depicted in Table 13.3 can be determined, which deserve the following clarification in the light of complexity:

- A vision of the project life cycle divided into two separate moments: project building and project driving
- A wider vision of the scope of project narration, which embraces the whole life cycle

It is self-evident that the implementation of traditional best practices in a complexity scenario cannot be limited to the adoption of one or more complementary approaches, such as the project systemic map or the emerging project charter. In fact it is necessary to be aware that a real interdisciplinarity of both approaches and models is the only way to improve the ability of working inside uncertain and rapidly changing contexts. For example, the adoption of an ethnographic approach, the use of informal networks analysis, and the ethical use of evaluation models for individual skills can all sensibly widen the skills that help to both interact and engage with the stakeholders, starting with the members of the project team.

A further question could arise from this comparison with traditional best practices: "To what extent can the outcomes of the research be applied to a wider context, for instance, program management, that is, a set of projects aiming at a common goal?" In order to answer such a complicated question thoroughly, a new research activity should be undertaken. In short, it could be said that the very nature of a program, characterized by the heterogeneity of the elements and by nonlinearity of corresponding relationships, is typical of a "complex adaptive system." Thinking about a program, it is easy to realize how essential it is to be able to operate in a systemic fashion, to promote distributed leadership, to get deeply in touch with the stakeholders' worlds, and to achieve a greater flexibility with respect to changes (maybe injecting some form of redundancy). That's why the topics tackled by this research are more evident and important in a wider context than in a smaller one.

Finally I would like to dwell on a more "pernicious" question which goes like this: "The topic of complexity has been widely discussed under different angles: what about the problem of how to measure complexity itself, which could be of great help in managing a project portfolio?" This is certainly a hot topic and, by the way, some tools are commercially available that can be used to estimate and measure the degree of complexity of the projects. That said, the idea that the Complexnauts have shared since the very beginning is that looking for a system of complexity indicators is just a shortcut, if not a wrong path, with respect to the real topic: to be able to grasp complexity variations within the project, putting in place the most effective actions to govern the evolution.

IT HASN'T FINISHED YET

The best way to end a book featuring so many authors is probably to quote an old Zen saying, which reminds us: "It's not the destination, it's the journey," or more explicitly, "The real destination is the journey." In fact it has been an authentic collective experience that lasted almost one year and a half (quite a long journey, isn't it?) where each Complexnaut has given his or her contribution in terms of both ideas and experiences, blending it all with three main ingredients: curiosity of exploring other disciplines searching for "food for thought," passion for a profession still capable of generating new stimuli for learning, and humility in offering potential solutions (and not imposing definite truths) to whoever is willing to try them and succeed using the same three ingredients.

This is our wish for the future and maybe the reason why we have tackled a practical topic offering both tools and approaches that complement the PMBoK. In fact we are all well aware that such approaches are not supported yet by significant evidence which can certify their strength and their reliability. At the same time, we think that this is an intellectually honest book, which concretely tries to answer the questions that complexity has asked us, using our own experience, and therefore our daily modus operandi (and, of course, our *modus cogitandi*).

A concluding promise: we are not going to measure the success of this book based on the number of copies sold, but rather based on two realities that we would like to see "emerge": first of all, the growth of a community of Complexnauts willing to continue this journey with us, devising and experimenting with new approaches; second, a gradual proliferation of a "humanistic" project management, strongly multidisciplinary and able to establish itself as the management of the future.

We really count on it!

REFERENCES

1. Varanini, F. (2008). *Complex Project Management: le ragioni di un percorso formativo-esperienziale*. Milan: Project Management Institute–Northern Italy Chapter.
2. De Toni, A.F. and Comello, L. (2005). Prede o ragni. In *Uomini e organizzazioni nella ragnatela della complessità*. Turin: UTET Libreria.
3. De Toni, A.F. and Comello, L. (2007). *Viaggio nella complessità*. Venice: Marsilio.
4. Bocchi, G. and Ceruti, M. (1986). *Il vincolo e la possibilità*. Milan: Feltrinelli.

Index

For Product Safety Concerns and Information please contact our EU
representative GPSR@taylorandfrancis.com
Taylor & Francis Verlag GmbH, Kaufingerstraße 24, 80331 München, Germany

www.ingramcontent.com/pod-product-compliance
Ingram Content Group UK Ltd.
Pitfield, Milton Keynes, MK11 3LW, UK
UKHW021621240425
457818UK00018B/671